Reinforced Concrete Structures – Innovations in Materials, Design and Analysis

Edited by Hosam M. Saleh,
Mohsen Mhadhbi and Amal I. Hassan

Published in London, United Kingdom

Reinforced Concrete Structures - Innovations in Materials, Design and Analysis
http://dx.doi.org/10.5772/intechopen.1001536
Edited by Hosam M. Saleh, Mohsen Mhadhbi and Amal I. Hassan

Contributors
Agnes H. Patty, Benedictus Sonny Yoedono, Sunik Sunik, Ján Staněk, Basel Elia Azar, Tomáš Fichtel, Emmanuel Ndububa, Winfred Nthuka Mutungi, Sanin Dzidic, Krystyna Rajczyk, Wiesław Kurdowski, Paweł Pichniarczyk, Grzegorz Janus, José Atílio Fritz Fidel Rocco, Rene Francisco Boschi Gonçalves, Marcela Galizia Domingues, Mustafa Kaya, Andrey Dolganov, Khanlar Seyfullaev, Yakubu Mustapha Karkarna, Ali Bahadori-Jahromi, Hamid Zolghadr Jahromi, Emily Halliwell, Musab Mohammad Rabi, Sérgio Orlando Antoun Netto, Lucas Pires Chagas Ferreira de Carvalho, Ana Waldila de Queiroz Ramiro Reis, Leonardo Vieira Barbalho, Lucas de Campos Rodrigues, Ali Goharrokhi

First published in London, United Kingdom, 2023 by IntechOpen
IntechOpen is the global imprint of INTECHOPEN LIMITED, registered in England and Wales, registration number: 11086078, 5 Princes Gate Court, London, SW7 2QJ, United Kingdom

British Library Cataloguing-in-Publication Data
A catalogue record for this book is available from the British Library

Additional hard and PDF copies can be obtained from orders@intechopen.com

Reinforced Concrete Structures - Innovations in Materials, Design and Analysis
Edited by Hosam M. Saleh, Mohsen Mhadhbi and Amal I. Hassan
p. cm.
Print ISBN 978-1-83769-493-8
Online ISBN 978-1-83769-492-1
eBook (PDF) ISBN 978-1-83769-494-5

For EU product safety concerns:
IN TECH d.o.o., Prolaz Marije Krucifikse Kozulić 3, 51000 Rijeka, Croatia,
info@intechopen.com or visit our website at intechopen.com.

Meet the editors

Hosam Saleh is a Professor of Radioactive Waste Management at the Radioisotope Department, Atomic Energy Authority, Egypt. He was awarded an MSc and Ph.D. in Physical Chemistry from Cairo University. Dr. Saleh has more than 26 years of experience in hazardous waste management with an emphasis on the treatment and development of new matrixes for the immobilization of these wastes. He is also interested in studying innovative economic and environmentally friendly techniques for the management of hazardous and radioactive wastes. He has authored many peer-reviewed scientific papers and chapters and served as an editor of several books. He was listed among the world's top 2% of scientists by Stanford University in 2020, 2021, and 2022.

Amal I. Hassan is a Professor of Animal Physiology, Department of Radioisotopes, Nuclear Research Center, Atomic Energy Authority, Egypt. She has authored many peer-reviewed papers on chronic disease problems. She is a referee, reviewer, and editorial board member for several international scientific journals. She received a Certificate of Excellence in international scientific research arbitration from Publons. She was also selected for inclusion in *Who's Who* in 2014, 2015, and 2016.

Dr. Mohsen Mhadhbi obtained his Ph.D. from the University of Sfax, Faculty of Sciences of Sfax, Tunisia. He is currently an assistant professor at the National Institute of Research and Physicochemical Analysis, Tunisia. His research interests include material engineering, ceramics, composites, energy, powder technology, simulation, and modeling. He has published numerous works in national and international journals and books. Dr. Mhadhbi has supervised several researchers in materials science. He is a member of various associations and an editorial board member for scientific journals.

Contents

Preface

The use of reinforced concrete has revolutionized the construction industry, allowing for stronger, more durable and versatile structures. Over the years, advancements in materials, design, and analysis have continued to push the boundaries of what is possible with this remarkable building material.

The collection of articles presented in this work delves deep into the various aspects of reinforced concrete, showcasing the latest innovations and cutting-edge research. From materials and properties to design and optimization, and even the identification of pathologies and the effects of corrosion, each section offers valuable insights into this fascinating field.

The first section, "Materials and Properties of Reinforced Concrete", explores the various materials and properties used in reinforced concrete, including cement-based materials, alternative and waste materials, fire resistance, and creep. The second section, "Design and Optimization of Reinforced Concrete", covers advanced design techniques and optimization algorithms, highlighting the challenges faced by design engineers and exploring new theoretical bases for the calculation of reinforced concrete elements. Finally, the third section, "Pathologies and Identification of Reinforced Concrete", presents studies on identifying pathologies in concrete structures, such as the application of terrestrial laser scanning, and the effect of corrosion on reinforced concrete columns.

The collection of chapters featured in this work is a testament to the ongoing dedication of researchers and practitioners in the field of reinforced concrete. It is our hope that this volume will serve as a valuable resource for those seeking to gain a deeper understanding of this important material and its many applications.

The editors wish to thank all the participants in this book for their valuable contributions. We also acknowledge the staff at IntechOpen, especially Author Service Manager Mr. Josip Knapic for his assistance in finalizing this work.

Hosam M. Saleh and Amal I. Hassan
Professor,
Radioisotope Department,
Egyptian Atomic Energy Authority,
Cairo, Egypt

Dr. Mohsen Mhadhbi
Assistant Professor,
National Institute of Research and Physicochemical Analysis,
Ariana, Tunisia

Section 1

Materials and Properties of Reinforced Concrete

Chapter 1

A Study on Toughness Contribution to Structural Capacity of Reinforced Concrete Beam

Agnes H. Patty, Benedictus Sonny Yoedono and Sunik Sunik

Abstract

The term of toughness is one of the fracture parameters which describes the ability of structures to remain deformed while collapsing. Toughness can be expressed as both, strain energy release rate G or as stress intensity factor K. This study deals with how reinforcement influences toughness K to divert rapid to gradual failure. Wedge forces developed by cohesiveness between rebars and concrete are the main concern in transforming elastic to plastic behavior by means reducing the value of stress intensity factor K. Three-point bend beam as a specimen with mode I fracture of (150 × 300) mm dimension with 100 mm initial crack was conducted in the analytical processing. The specimen was reinforced by 4#12mm steel bars. Wedge forces '*p*' due to reinforcement tensile T developed by composite action between concrete and reinforcement prevailed $K_I^P = 441.613$ Nmm$^{-1.5}$ whereas stress intensity factor due to load for beam without reinforcement $K_I^P = 482.7$ Nmm$^{-1.5}$. Hence, the stress intensity factor due to the contribution of reinforcement $K_I^R = 41.087$ Nmm$^{-1.5}$ which is greater than the critical toughness $K_{Ic} = 22.136$ Nmm$^{-1.5}$. By applying the term strain energy release rate G in conjunction with stress intensity factor K through the relationship $K = \sqrt{EG}$, resulting in G is close to 35 N/m, a value under which normal plain concrete would fail.

Keywords: fracture mechanics, toughness, stress intensity factor, wedge forces, reinforced concrete beam

1. Introduction

Concrete is well known as discrete material which is full of cracks. This condition occurs due to some reason, such as imperfection during the mixing of its constituent, casting and curing process as well [1].

This unfortunate initial condition prevails to decrease the ability of toughness performance after elastic limit is achieved. Cementitious material such as concrete with relatively high discretization is weak to the separation (fracturing) between the constituent grains. Therefore, in assessing its structural capacity, it needs to be specially formulated by taking into account the possibility of developing fracture process zone - FPZ at the crack tip through toughness vice versa [2–4]. The development of material technology has brought the 'world of science' to develop up-to-date concepts related to effective and efficient structural capacities which mean that a structural design must fulfill safe and economical aspects. Applying this idea, structural failure should be perceived based on both strength criteria and fracture mechanics view of point.

Mathematically, failure mechanism which is based on the principles of fracture mechanics can be expressed as total potential energy, that is $\Pi = U - F + W$, where U is strain energy of the structure, F is the work done by the applied force, and W is the energy needed for crack formation [5]. In the case of reinforced concrete, the reinforcement significantly through bonding forms a fracture process zone (FPZ) at the crack tip. This FPZ acts as a crack arrester which in turn increases toughness [6]. T - in **Figure 1**, is the tensile force under balanced condition at which the reinforcing steel and the compressive concrete simultaneously reach their plastic moment M_P as their ultimate capacities.

2. Review

2.1 Toughness

Structures under loading will undergo two basic stages of deformation, namely 'elastic' and 'plastic'. Elastic is a phenomenon where deformation will return to its original shape when load is removed. While plastic is post elastic deformation, yielding or semi yielding or quasi-brittle depends on material characteristic. To get a clear picture of these two phenomena, consider the load deformation relationship in **Figure 2** [7].

The area up to the yield point is known as resilience or elastic range, while area beyond the elastic limit up to ultimate point at which failure occurs is known as toughness. Based on the concept of fracture mechanics, resilience describes energy consumed during loading, but toughness is energy dissipated while fracture propagation in progress/occurs.

In addition, toughness can be analogous to ductility in yielding materials.

2.2 Structural collapse capacity

Structural collapse can be defined as a condition where structures can no longer able to withstand load. This phenomenon is the starting point on which capacity design is based. Referring to the concepts of fracture mechanics, structural collapse

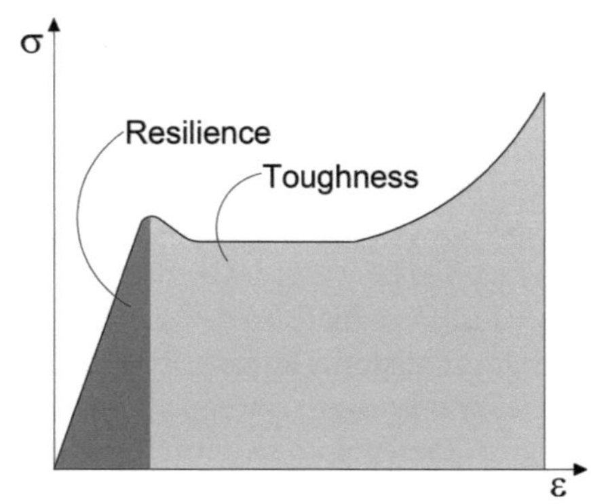

Figure 1.
Wedge forces of a single crack reinforced beam.

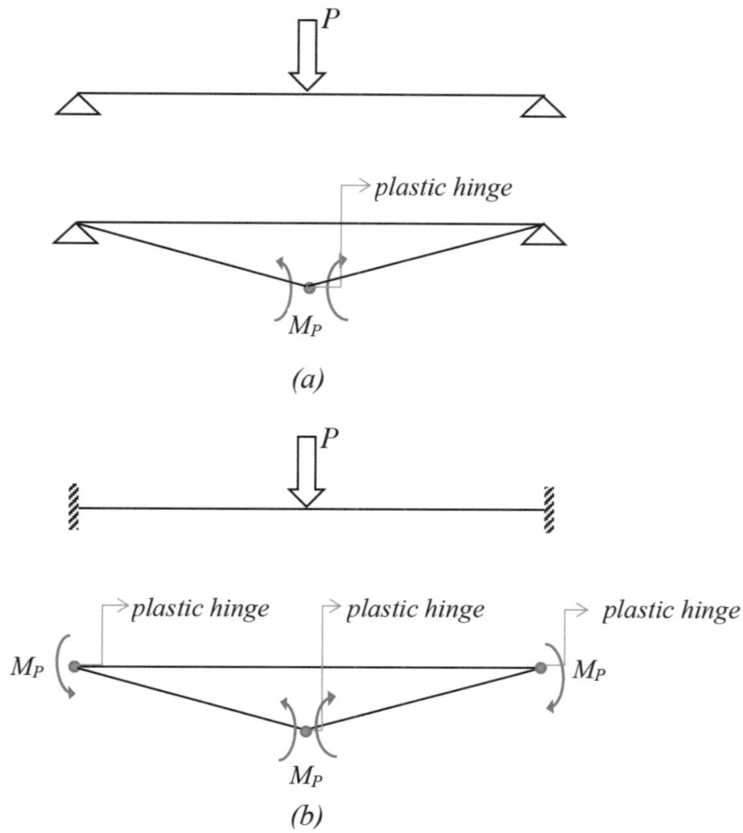

Figure 2.
Resilience vs Toughness [7].

capacity can be assessed by studying the possibility of FPZ propagation (at sectional level) followed by the formation of plastic hinges (at structural level). Please keep in mind that fracture zone is a term of damage zone performed at the crack tip when a beam is loaded beyond its elastic limit [8]. By considering the illustration in **Figure 3**, it can be proved analytically that a beam with free rotation supports shown in **Figure 3a** does not have toughness as high as a beam with fixed support shown in **Figure 3b**. Plastic hinges are components which enable to dissipate energy as fracture propagation occurs. This phenomenon is potential to divert brittle fracture caused by the absence of toughness into a gradual failure due to the presence of toughness.

Toughness index rules the performance of collapse mechanism. Resilience refers to 'strength' in the elastic range before collapse, while toughness refers to the ability of structures to 'deform' before collapse.

2.3 Stress intensity factor

The term 'stress intensity factor SIF is one of the fracture parameters; it describes stress intensity around the crack tip, which is represented by variable K. Depending on its critical value K_c, structural failure capacity might be analyzed based on the viewpoint of fracture mechanics. When K due to loads exceeds the value of K_c, then fracture failure is unavoidable. Concrete structures are known to be full of cracks; this

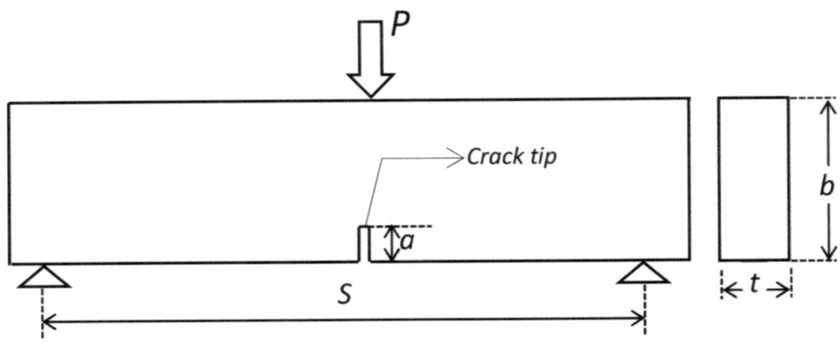

Figure 3.
Illustration, structural collapse capacity described by plastic moment M_P (a) Beam with end free rotation supports – 1 plastic moment M_P. (b) Beam with end fix support – 3 plastic moment M_P.

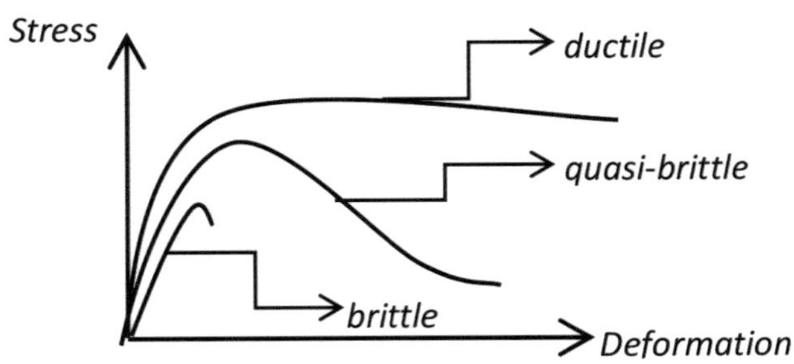

Figure 4
Three point mode I fracture beam.

phenomenon comes from initial internal discontinuity which occurs due to some reasons, such as imperfection during the mixing of constituent component of the material. This internal crack has the potential to increase the stress intensity factor *K*, around the crack tip.

Developed in 1957 by Irwin [9], *K* defines the magnitude of local stress around the crack tip. Please note that this factor depends on loading, crack size, crack shape and geometric boundaries, with the general form given by

$$K = \sigma\sqrt{\pi a}\, f(g) \tag{1}$$

where σ = remote stress applied to the component; a = crack length; f(g) = correction factor that depends on specimen and crack geometry.

In case of three-point bend beam shown by **Figure 4**, Eq. (1) can be re-written as [5]

$$K_I = \sigma\sqrt{\pi a}\, g_1\left(\frac{a}{b}\right) \tag{2}$$

where a = length of initial crack; b = depth of beam; t = thick of beam; $g_1\left(\frac{a}{b}\right)$ = is geometry function, calculated as

$$g_1\left(\frac{a}{b}\right) = \frac{1.0 - 2.5\left(\frac{a}{b}\right) + 4.49\left(\frac{a}{b}\right)^2 - 3.98\left(\frac{a}{b}\right)^3 + 1.33\left(\frac{a}{b}\right)^4}{\left(1 - \frac{a}{b}\right)^{3/2}} \tag{3}$$

The remote stress σ is nominal normal stress due to P, acts perpendicular the crack face, counted as [5]

$$\sigma = \frac{3PS}{2tb^2} \tag{4}$$

To avoid structures from brittle fracture, the value of K_I must be below its critical value K_{Ic}. Brittle fracture can be described as unstable crack growth. In case of reinforced concrete beam, the stress intensity factor K due to applied load can be reduced by traction contributed by bridging effect between reinforcement and the concrete is self.

2.4 How steel bars reducing *K*

In principle, failure of concrete begins with the collapse of the material due to its inability to expand the composite action between matrix and aggregates. It is highly recommended to understand that the interface zone rules a role in such a case. In plain concrete, aggregates have the potential to function as reinforcement which reinforces matrix. Steel bars play the same role at higher level, namely reinforced concrete. Due to steel bars reinforcement, the accumulated stress intensity at the crack tip can be reduced prevails a stable fracture failure mechanism. This is what the concept of ductility underlies especially in earthquake-resistant structures. Reinforcement is a driving force that generates composite action between rebars and concrete, and hence, it is able to divert the brittle failure gradually to quasi-brittle and finally to ductile, as shown in **Figure 5**.

2.5 Fracture toughness of concrete

Fracture toughness can be clarified as both, critical strain energy release rate denoted by G_C and critical stress intensity factor denoted by parameter K_C. These two parameters meet the following relationship

$$K_c = \sqrt{EG_c} \tag{5}$$

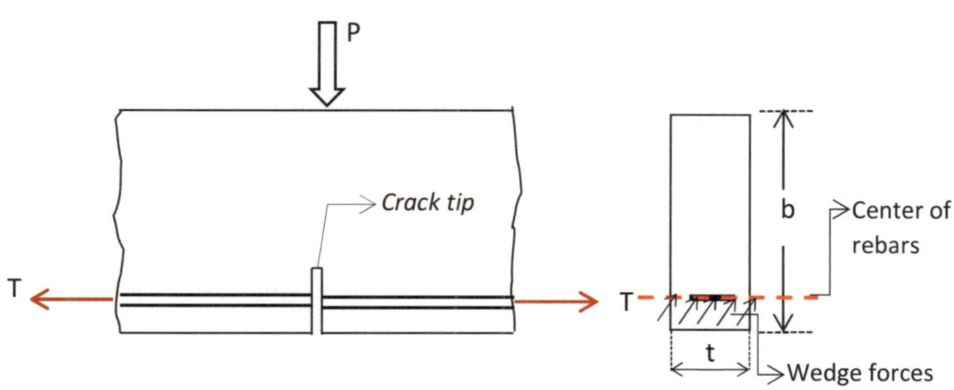

Figure 5.
Wedge forces of a single crack reinforced beam.

Concrete is considered as a quasi-brittle material whose failure mechanism might be expressed by the compatibility between the energy absorbed when loading and energy released when unloading. Please be noticed that in case of crack propagation, unloading occurs simultaneously with crack propagation for quasi-brittle materials, analog with yielding process for ductile materials [3, 10].

2.6 Wedge forces of single cracked element

Concrete is a material with a high sensitivity to notches. If it is subjected to a tensile force, e.g., bending reinforced concrete beam shown in **Figure 1**, then cracking propagation will occur and that starts from the crack tip at the moment when tensile stress due to loading exceeds its tensile strength.

In case of under-reinforced or over-reinforced concrete beam one can visualize an extreme mode of failure known as brittle fracture mechanism which has to be avoided especially in case of dynamic loads. Reinforced concrete with adequate reinforcement ratio is able to develop wedge forces due to composite action between reinforcement and concrete. Under such condition, beam will collapse slowly as the fracture process forms at the crack tip ... known as being tough [1, 3, 11]. This study examines how reinforcement reduced the stress intensity factor SIF and simultaneously increase toughness.

3. Case study

A reinforced concrete beam of (150 $\times$ 300) mm^2 dimension with span S = 1000 mm was subjected to a centralized load P at the middle span. A beam with a single tensile crack was made of concrete material with properties given as follows [7]

Compressive strength

$$f_c' = 104.30 \text{ MPa}$$

Fracture toughness

$$K_{Ic} = 22.136 \text{ N.mm}^{-3/2}$$

Modulus of elasticity of steel bar

$$E = 48{,}000 \text{ MPa}$$

Yield strength of steel bar

$$f_{ys} = 590 \text{ MPa}$$

If the initial crack a = 100 mm, with 4$\neq$12 mm reinforcement, this study was conducted to investigate whether the reinforcement is sufficient to reduce the stress intensity factor KI at the crack tip so, the brittle failure can be diverted to quasi-brittle or even to ductile/plastic failure. Please note that K_I is stress intensity factor for mode I fracture beam. Referring to **Figure 6**, the chronological analysis is as follows.

3.1 Step I: checking the reinforcement ratio

T is the tensile force developed by the reinforcement. The minimum amount of reinforcement should be provided such that the first cracking of concrete and yielding of steel occurs simultaneously [11].

Bosco et al proposed a brittleness number corresponding to a transition between ductile ($\rho > \rho_{min}$) and brittle failure ($\rho < \rho_{min}$) that may be expressed as a function of compressive strength of concrete as

$$\rho_{min} = (3.1 + 0.0023 fc') \frac{K_{Ic}}{f_y b^{\frac{1}{2}}} \tag{6}$$

found, $\rho_{min} = 0.000131$

Maximum ratio may be calculated using the basic formula as

$$\rho_{max} = 0.75 \left(\frac{0.85 f'_c \beta_1}{f_y} \right) \times \left(\frac{600}{600 + f_y} \right) \tag{7}$$

found, $\rho_{max} = 0.036934$

The results shown that $\rho_{min} < \rho < \rho_{max}$ which condition prevails to a ductile mechanism of failure.

3.2 Step II: determination of neutral axis

By assuming that plane section remains plane after deformation, a linear distribution of strain over the beam depth was obtained.

Neutral axis is determined based on the linear strain relationship in **Figure 6**.

If the ultimate strain of concrete $\varepsilon_{cu} = 0.002$, prevailed $c = 101.67$ mm.

3.3 Step III: determination of moment capacity M_{nf}

1. Force equilibrium

2. $C = 0.85 f'_c (\mathrm{a})(\mathrm{t})$

   T = $Asfy$ = 266774.4 N

   z = $h - \frac{a}{2}$ = 233.97 mm

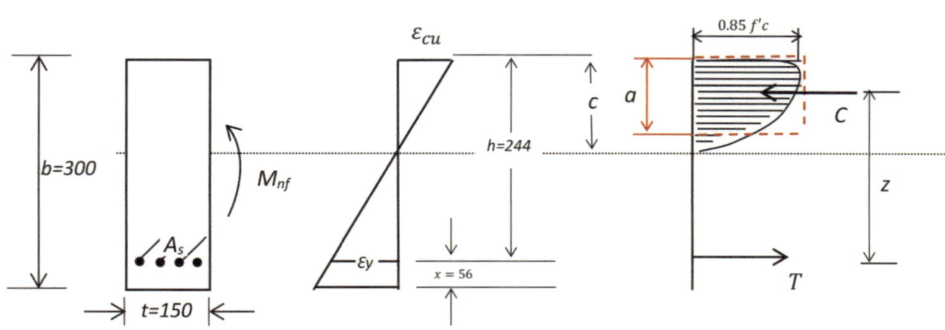

Figure 6.
Strain and stress distribution under bending.

3. M_{nf} = T or C x (z)

= 62417206.62 Nmm

3.4 Step IV: determination of stress intensity factor due to P

1. Remote stress due to P [5]

$$\sigma = \frac{1.5\,PS}{b^2 t}$$

or,

$$\sigma = \frac{M}{\frac{1}{6}(150)(300)^2}$$

Applied to ultimate condition where $M = M_{nf}$ resulted in

$$\sigma = 27.74\ N/\text{mm}^2$$

1. $g_1\left(\frac{a}{b}\right) = \frac{1.0-2.5\frac{a}{b}+4.49\left(\frac{a}{b}\right)^2-3.98\left(\frac{a}{b}\right)^3+1.33\left(\frac{a}{b}\right)^4}{\left(1-\frac{a}{b}\right)^{3/2}}$ = 0.982

2. $K_I^P = 27.74\sqrt{\pi.100}(0.982) = 482.70\ Nmm^{-3/2}$

3.5 Step V: toughness contributed by reinforcement

1. Wedge forces $p = \frac{T}{t}$ where T is tensile force acts to rebars due to bending.

2. Refering to Green Function, stress intencity factor due to average stress over the crack line [12]

$$K_I = \frac{2p}{\sqrt{\pi a}} g_1\left(\frac{a}{b}, \frac{x}{a}\right) \ldots . \quad (8)$$

here, x = 56 mm was the distance between center of reinforcement and the outer tension fiber of the beam.

$$g_1\left(\frac{a}{b}, \frac{x}{a}\right) = \frac{3.52\left(1-\frac{x}{a}\right)}{\left(1-\frac{a}{b}\right)^{\frac{3}{2}}} - \frac{4.35-5.28\frac{x}{a}}{\left(1-\frac{a}{b}\right)^{\frac{1}{2}}} + \left[\frac{1.30-0.30\left(\frac{x}{a}\right)^{\frac{1}{2}}}{\sqrt{1-\left(\frac{x}{a}\right)^2}} + 0.83 - 1.76\frac{x}{a}\right]\left[1-\left(\frac{x}{a}\right)\right)\frac{a}{b}\right] \quad (9)$$

found, $g_1, \left(\frac{a}{b} - \frac{x}{a}\right) = 2.2$

Wedge forces due to reinforcement T was:

$$p = \frac{266774.4}{150}$$

Then, $K_I^P = \frac{2p}{\sqrt{\pi a}} g_1\left(\frac{a}{b}, \frac{x}{a}\right) = 441.613 \text{ N.mm}^{-3/2}$

3. Reinforcement contribution to K_I equals to $K_{Ic}^R = K_I^P - K_I^p = (482.70 - 441.613)\text{Nmm}^{-3/2} = 41.087\text{Nmm}^{-3/2}$

4. Discussion

1. Resume of analysis results

    - The critical stress intensity factor, K_{Ic}= 22.136 $\text{Nmm}^{-\frac{3}{2}}$
    - The stress intensity factor due to wedge forces T, $K_I^p = 441.613 \text{ Nmm}^{-\frac{3}{2}}$
    - The stress intensity factor due to the external load P, $K_I^P = 482.70 \text{ Nmm}^{-\frac{3}{2}}$
    - Based on the principle of total potential energy [5], the stress intensity factor due to the contribution of reinforcement is equal to 41.87 $\text{Nmm}^{-\frac{3}{2}}$.

2. Without reinforcement, stress intensity factor $K_I^P = 482.70 \text{ Nmm}^{-3/2}$ is much higher than K_{Ic}=22.13 $\text{Nmm}^{-3/2}$ means that crack will run rapidly as common as in high strength concrete. To avoid this phenomenon high performance concrete stuctures must be designed using the non-linier fracture mechanics, in this case contributed by toughness due to the presence of reinforcement [13].

3. The results of the analysis shown that due to reinforcement presence, crack propagation still occurs under a value of stress intensity factor K_I^R higher from the critical value K_{Ic}.

4. This fact implies that brittle failure dealing with K_{Ic} occurs to concrete, even to high strength, can be diverted to quasi-brittle/semi-plastic, and even to plastic failure due to the presence of reinforcement.

5. Conclusion

- Concrete as a discrete material tends to be full of cracks leading to rapid failure mechanism.
- Rapid failure mechanism can be diverted to quasi-brittle mechanism by performing fracture process zone FPZ, at the crack tip, resulting to softening type of load-deformation relationship.
- By adding reinforcement, the FPZ will become more solid resulting in toughness improvement and energy dissipation.
- Toughness is one of the most prominent fracture parameters on which structural capacity can be defined.

Author details

Agnes H. Patty, Benedictus Sonny Yoedono* and Sunik Sunik
Widya Karya Catholic University, Malang, Indonesia

*Address all correspondence to: sonny_ft@widyakarya.ac.id

References

[1] Leonhardt F. Crack and crack control in concrete structures. PCI Journal. 1988; **33**(4):124-145

[2] Rama Chandra Murthy A, Palani GS, Iyer NR. State of the art review on fracture analysis of concrete structural components. 2009; 345–367

[3] Patty AH. The effect of reinforcement bridging on the elastic fracture energy of concrete. Material Science and Engineering. 2019;**669**. DOI: 10.1088/1757-899X/669/1/012019. (Open access file)

[4] Bhowmik S, Ray S. An experimental approach for characterization of fracture process zone in concrete. Engineering Fractures Mechanics. 2019;**2019**:401-419. DOI: 10.1016/engfreacmech.2019.02.026

[5] Shah SP, Swartz SE, Ouyang C. Fracture Mechanics of Concrete. USA: John Wiley & Sons, Inc; 1995

[6] ACI Committee 446.1R-91. Fracture mechanics of concrete: Concepts, models and determination of material properties. 1989

[7] Roylance D. Mechanics of Material. USA: John Wiley & Sons, Inc; 1996

[8] Hoover CG, Bazant ZP, Wendner R, Vorel J, Hubler MH, Kim K, et al., editors. Experimental Investigation of Transitional Size Effect and Crack Length in Concrete Fracture: Life-Cycle and Sustainability of Civil Infrastructure Systems. London: Taylor & Francis Group; 2013

[9] Irwin GR. Analysis of stresses and strains near the end of a crack traversing a place. Journal of Applied Mechanics. 1957;**24**:361-364

[10] Zdenek PB. Concrete fracture models: Testing and practice. Engineering Fractures Mechanics. 2022; **2022**:165-205

[11] Bosco C, Carpinteri A, Member ASCE, Debernardi PG. Minimum reinforcement in high strength concrete. Journal of Structural Engineering. 1990; **1990**:427-437

[12] Cartwright DJ, Rooke DP. Green Functions in Fracture Mechanics. Future Prospects, University of Cambridge; 1979

[13] Ricardo AE, Velasco MSL. Fracture parameters for high-performance concrete. Cement and Concrete Research. 2006;**36**:576-583. DOI: 10.1016/j.cemconres.2005.09.004

Chapter 2

Cement-Based Materials in Dentistry

Ján Staněk, Basel Elia Azar and Tomáš Fichtel

Abstract

Cement-based materials in dentistry have experienced rapid development. In the field of operative dentistry, there are mainly developing calcium silicate cements, which have made it possible to solve previously difficult situations such as perforation of the root-canal system, direct pulp capping, filling and preserving teeth with widely open foramen apicale. These materials are based on the Portland cement. This chapter will describe the development, properties, indications and limitations of these materials. In the field of prosthodnotics, the prosthetic restoration is connected to the remaining tissues with the help of cements. Requirements for such materials and the available options will be described. The choice of suitable cement is based on its properties, requirements (such as moisture control), the material of the restoration (optimal choice can affect and strengthen the material) and the characteristics of the remaining dental tissues (such as the conicity of the prepared tooth). The chemical preparation of the tooth and prosthetic material connected with the individual types of the cements, which are capable to ensure the firm connection leading to the long-term and aesthetic result, will be described.

Keywords: operative dentistry, prosthodontics, calcium silicate-based cements, water-based cements, adhesive cementation

1. Introduction

According to ISO 9917-2:2017, cement-based materials in dentistry can be used as luting materials, as base or lining material, as material of restoration and as tooth core build-up [1].

The aim of this chapter is to present new treatment options and considerations based on up to date research and knowledge about cement-based materials in dentistry. For didactical and comparative reasons, the material with long history of use is discribed, too.

In operative dentistry, cements are used as base or lining material and as material of the direct provisional restoration.

Calcium silicate cements are the most developing materials in operative dentistry for their biological proprieties. Recently, the interaction of the calcium silicate materials and stem cells from periodontal ligament has been studied in vitro. The calcium silicate cements showed adeqaute osteogenic and cementogenic potential.

The influence of additives, mixing technique and different condition on biological proprieties and clinical success stays as a question for future [2].

As recent research studies showed with use of calcium silicate cements immature teeth with pulp inflamation can be maintained and even the development of the tooth can continue, when the diagnosis is done properly, biology and material are understood and the protocol for treatment is done properly [3].

According to up-to-date proofs, the pulpotomy can be considered as less invasive alternative to root canal treatment for teeth with massive caries destruction and with pulp alteration, even if it is irreversible in the defined part of the pulp. Consideration for treatment and material selection are going to be described. New materials allow clinican less invasive treatment options [4].

For its high biokompatibility, the calcium silicate materials are also studied apart from dentistry as a drug delivery system [5].

Cement-based materials in prosthodontics are used for luting indirectly fabricated restoration and core build-up.

Composite resin-based cements (adhesive cements) are indicated for core build-up for their mechanical proprietes.

Luting could be done as provisional or definitive.

The connection of prosthesis and dental tissue—cementation is an extremely important step in prosthodontics. The prosthesis is cemented to different prosthetic materials, different dental tissues and filling materials. Individual materials have different requirements for maintaining dryness in the oral cavity. Based on material characteristcs, the suitable cement is choosen. Materials of the cementation are constantly evolving and improving.

As in 2019, there was a question if universal adhesives can bond to zirkonium ceramics, as this is unpretreatable by etching in contrast to ceramic with glass, nowdays the protocols for adhesive cementation are defined and can be discussed [6].

New protocols and molecule of 10-Methacryloyloxydecyl dihydrogen phosphate (10-MDP) allow adhesive fixation of zirconium ceramics, what used to be unpredictable and challenging. This presents a milestone in a prosthodontic dentistry. The higher clinical success of ceramic restorations is expected [7].

However, the adhesive layer can be decomposed by enzymes matrix metalloproteinases, what makes it instable in time. Recent studies show that the molecules such as chlorhexidine digluconate and captopril are able to inhibit these enzymes. Effectivity of these molecules and other alternatives should be researched [8, 9].

Improper selection of cementation material, incorrect use or imperfect removal of excess can result in premature failure.

Cement ensures the retention of the restoration. In the following chapter, the protocoles and materials are discussed as well as materials with long history of use. This chapter goes systematicly through different options of cements use and summarises evidence-based protocols, considerations for modern materials selection and discusses their advantages and disadvantages.

2. Cements in operative dentistry

2.1 Provisional restoration materials

Provisional filling is used for two main reasons: for sealing endodontic access before final root canal obturation to prevent reinfection, and from time reasons to maintain esthetics and functionality of teeth which will be repaired by definitive filling in near future.

Regarding deep caries lession, the marginal seal could not be suficient. It is clinicans' choice to restore with provisional filling and wait or with definite one which can also act as preendodontic build-up in case of failure.

Different materials can be used such as zinc polycarboxylate (also known as zinc polyacrylate), zinc phosphate cement, zinc oxid sulphate and glass ionomer cement (for filling material maximum particle size is 50 um, mechanical proprieties are described American National Standards Institute–American Dental Association (ANSI-ADA) Specification No. 96 (ISO 9917 [2000]). Although the glass ionomer releases flourides, it is sensitive during setting and the final the surface can be rough, so the author does not see an advantage for patients with low oral hygiene level [10, 11].

2.2 Base and lining material

Those are materials placed between restoration and dentin with the intention to protect the pulp. They were used espcialy under amalgam fillings.

Dental liner is placed in thinner layer, then dental base material.

Liner is a material that should idealy seal dentinal tubules and protect the pulp.

As liner, a varnish and calcium hydroxide can be used. In literature, glass ionomer and resin can be found, but these materials are used as a whole filling.

Layer of liner is approximately 0.5 mm.

Varnish is natural gum dissolved in solveont which is going to evaporate after its placement to cavity, leaving thin layer, thus reducing microleakage. It was used historicaly with amalgams. Varnish is wash-out of the margin and inhibits bonding of glass ionomer [12].

Calcium hydroxide has ability to stimulate tertiary dentin formation. It can be used even for direct pulp capping. Calcium hydroxide can be supplied in two paste forms or as light-cured material. It shows bactericidal proprieties, pH is around 11–12, and it is highly soluble. The material is moisture sensitive with weak mechanical proprietes. It can be used in proximity to pulp approximately 1–2 mm. Calcium hydroxide can be placed under the base material. There is a long history of use of calcium hydroxide. Calcium hydroxide can be also used in endodontics as intracanal medicament to desinfect root canal system [13].

Base materials are zinc oxid phosphate, zinc oxid eugenol, zinc polycarboxylate (zinc polyacrylate). Dental base material should protect the pulp by thermal isolation and by absorbing oclusal forces. Dental specialist can find these materials after removing old amalgam restoration, where they were used.

Glass ionomer and composite resin are not used as base materilas, but they have the same characteristcs such as providing barrier. Composite resin can be shaped and prepared complex morphology of the tooth surface for indirect restoration [12].

2.3 Calcium silicate cements and its development

Calcium silicate cements are hydraulic cements based on portland cement. First commercialy available product was Pro Root MTA (mineral trioxide aggregate). In literature, sometimes MTA is used for whole group of cements—calcium silicate cements. Calcium silicate materials are setting by hydration, and they have low solutibility. During setting the cements are slightly expanding in volume.

Original patent for MTA—the first of a large generation of calcium silicate cements—"MTA consists of 50-75% calcium oxide and 15-25% silicon dioxide." MTA is type 1 portland cement according to American Society for testing and materials Standard C150/C150M–12 2012–12 2012. Bismuth oxide was added to the first product MTA as radioopaficer [14].

Clinker of portland cement consists mainly of tricalcium silicate. Dicalcium silicate reacts and solidifies slower. These two components can be 80% of potland cement. Around 10% is gypsum. Calcium carbonate forms relatively bigger particles, thus reacting as nuclaeting centrum, thus accelerating setting. As 3–4 h are clinically in some situations unacceptable. Calcium carbonate is contained in Biodentin. This material sets in 12 minutes, and it can be used as provisional filling. Dicalciumphosphate is a dash of the latest calcium silicate materials, which encourages the production of hydroxypatite layer. This can lead to higher bond to dentin. Dicalciumphosphate is a dash of the latest calcium silicate materials, which encourages the production of hydroxypatite layer. This can lead to higher bond to dentin [15, 16].

Calcium silicate materials in wet condition release Ca2+ ions, these react with phospahtes present in the blood, and on the surface of calcium silicate material, the apatite/hydroxyapatite layer is formed. This is the reason of the biological tolerance of this material [17].

MTA sets (medium setting time) at 165 + −5 min. It reacts with water [18].

The pH after mixing was measured to be 10.2. It rises to 12.5 in 3 h and then stays constant [19].

During mixing, the higher volume of fluids leads to increased porosity and consequent solubility [20].

Calcium silikate materials unlike calcium hydroxide are not absorbed so quickly and lead predictably to form an apical seal. This is used for obturation root canals with wide apical foramen [21].

As mentioned before, the MTA expands slightly during solidification, which contributes to marginal seal. This depends on the proportion of water contained in the mixing.

Gypsum is added to MTA to influence time of the setting. Bi_2O_3 was added to increase radiopacity. Bi_2O_3 after contact with NaClO (commonly used root canal disinficient) changed its colour to dark brown. Radioopaficier can migrate in dentin and cause tooth disoloration. Other radioaque matherials are used in different calcium silicate cements such as zirconium dioxide or tantalum pentoxide [22].

To facilitate manipulation in latest materials, propylenglycol can be used by the manufacturer to adjust consistency [23].

Accelerators of setting such as calcium chloride can be used, what was a new step of the material development. This allowed to introduct provisional filling material based on calcium silicate cements. Material called Biodentin (Septodont) sets 12 min, what is clinically acceptable time of setting. This material is indicated for direct and indirect pulp capping [24].

Calcium silicate cements can be mixed by dental assistent or mechanically. They can be delivered premixed in syringe or as powder and liquide or as paste in consistency. They can be delivered unpremixed such as powder and liquid. Diffrent consistencies are available such as sealer, putty, paste.

Delivery on its place is possible by plugger, map system, dovgan applicator, microcarrier, or Lentulo spiral [25–29].

Development of these materials was follow these principles. Adding radioopaficier. And subsequently searching for radioopaficier which does not react with natrium hypochloride and does not cause tooth discolouration. Easier preparation and easier handling were achieved. Influencing setting time and thus allowing easier clinical application.

Clinical applications of the calcium silicate cements are pulp capping (direct or indirect), pulpotomy, open apex obturation, root perforation repair, endodontic apical microsurgery, external and internal resoption repair and regenerative endodontics and as a root canal canal sealer.

2.3.1 Direct pulp capping

During treatment of deep caries, in dental traumatology or even during preparation for prosthetic work, the vital pulp can be exposed. To avoid inflamatory complications in dental pulp, this tissue should be treated in correct way (**Figure 1**).

MTA was found as a superior material for a **direct pulp capping** when compared with CaOH in research with 2-year follow up [30].

Direct pulp capping for success requires uninfected pulp without inflamatory changes and peripheral seal which does not allow further bacterial iritation. Differential diagnosis between reversible and ireversible changes in dental pulp cannot be based only on patient complaint, but more objectively is based on ability to ensure hemostasis during 5–10 min with 1.5–6% sodium hypochlorite. Calcium silicate cement used for direct pulp capping should not contain bismuthtrioxide because of discolouration risk. Calcium silicate cement should be applied through the preforatin to ensure proper bond to hard dental tissues. The antimicrobial aktivity of calcium silicate material is

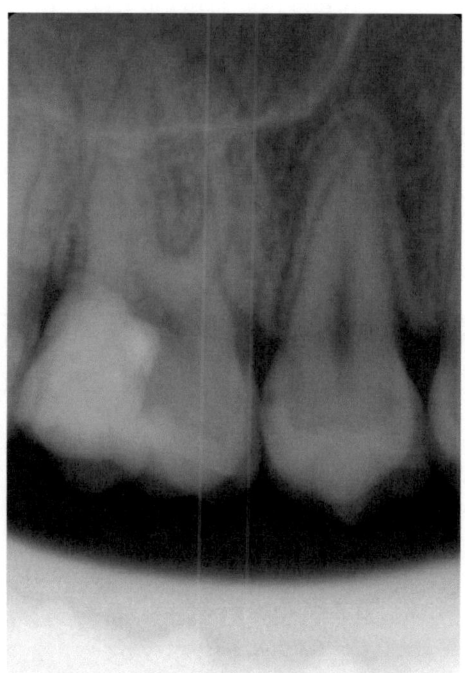

Figure 1.
Large carious defect resotred by composite filling, in the pulp chamber the calcium silicate cement used for direct pulp capping is visible. Three years after treatment the reaction on the cold test is positive.

advantageous. Clinacaly reasonable is utilisation of materials with shorter setting and with higher compressive strength resistence [31, 32].

2.3.2 Pulpotomy

Pulpotomy according to American Association of Endodontists means removing irreverisibly inflamatory changed coronal part of the pulp and remaining and preserving radicular part of the pulp. Pulpotomy can be done on permanent teeth with or without closed apical foramen (what is a sign of the end of the root development). Pulpotomy can be also done on decidual teeth. Procedure is done by sterile diamond bur or by electorsurgery. After hemostatis the CaOH or calcium silicate cement is placed and the tooth is adhesively reconstructed. Calcium silicate cements are superior materials in this indication. Logically it is not indicated when whole pulp is necrotic and apical periodontitis is present. The vitilaty tests and finding pulp tissues intracoronaly are the mandatory conditions. Periapical radiographs can mislead clinican by periapical rarefaction—flase-positive presence of periapcal radiolucency. Pulpotomy of permanent teeth with closed foramen apicale is advantageous when the complicated root canal anatomy is present, also from time reasons and requirements for minimal invasivity. When the hemostasis cannot be obtained, the conventional endodontic treetment is indicated.

The pulpotomy on vital undeveloped teeth with pulp exposure is justified by intent to end the root development.

Pulpotomy on decidous teeth is technically very difficult, because children as patients are not so cooperative as adult patients and handling with rubberdam can be very difficult. General anestesia can be the solution for uncooperative patient, but its indications should be judged carefully [33, 34] (**Figures 2** and **3**).

2.3.3 Open apex obturation

Open apex occurs after pulp necrosis at incomplete root development what may be caused by dental trauma, iatrogenic by overinstrumentaion and as result of inflammatory root apex resorption when the root canal infection is present. Thin walls of immature teeth with necrotic pulp are more prone to fracture,

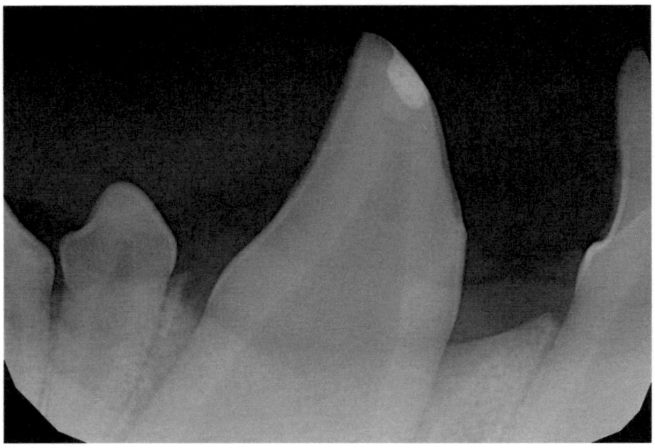

Figure 2.
Dental trauma in dog. Pulpotomy was performed using calcium silicate cement and composite filling.

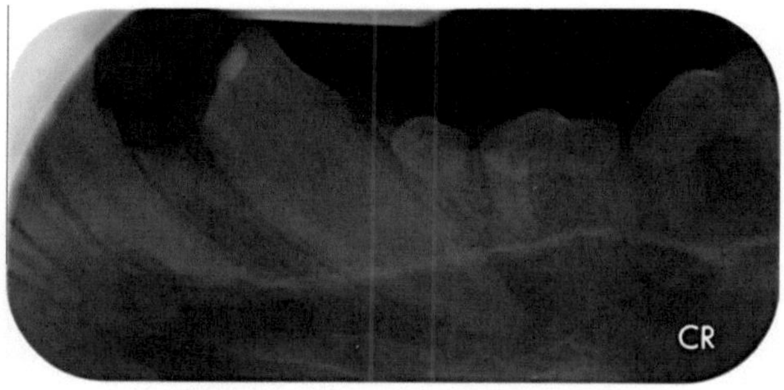

Figure 3.
Dental trauma in dog. The pulp tissue stays vital after pulpotomy using calcium silicate cement and composite filling.

additional layer of calcium silicate material at least 3 mm thickness should be placed. Thin walls of immature teeth with necrotic pulp are more prone to fracture, additionally. The clinical challenges are in the determination of working length as well as optimal obturation avoiding the extrusion of the material beyond foramen apicale.

First of all, the appropriet approach should be choosen. The clinican should choose if it is desirable to create "apical constriction". Apexification is the procudure based on repative aplication of calcium hydroxide, thus creating calcified aoical barier. This can take from 5 to 20 months.

Due to development of calcium silicate materials, single visit obturation of open apex is now possible.

A layer of calcium silicate material at least 3 mm thickness should be placed. Can be condensed against the inner matrix for resorbable hemostipic material which does not allow extrusion through the apex. Than the adhesive coronal seal is made (**Figures 4–6**) [35].

Regenerative endodontics is another option to create apical stop using calcium silicate materials placed on blood clot. Although the changes in root apex are visible on radiographs, the dentin in the cervical area will not regenerate. Histological finndings and practical consideration are discussed in the chapter on regenerative endodontics.

2.3.4 Root perforation repair

Success of treatment of root perforation repair is affected by localisation of perforation, abscence of contamination and time to the definitive treatment. Root perforation contaminated by communication with oral cavity (for example, by perio pocket) has the worse prognosis for healing. The higher the perforation is, the mechanicaly weaker the tooth is. Bleeding control is more difficult, so the inner matrix technique can be the option. Root perforation in supracrestal area is reparable by composite resin with combination with surgical approach. The success with calcium silicate cement is jeopardised by wash-out. Calcium silicate cements should not block visibility or make it impossible to find and work in root canals. This is the reason why the calcium silicate material should be placed after root canals negotiation. Apical perforation can be treated by ortograde access or retrograde microsurgical access. Coronal perforation is in danger of periopocket formation, even after surgical repair of the perforation (**Figure 7**) [36–38].

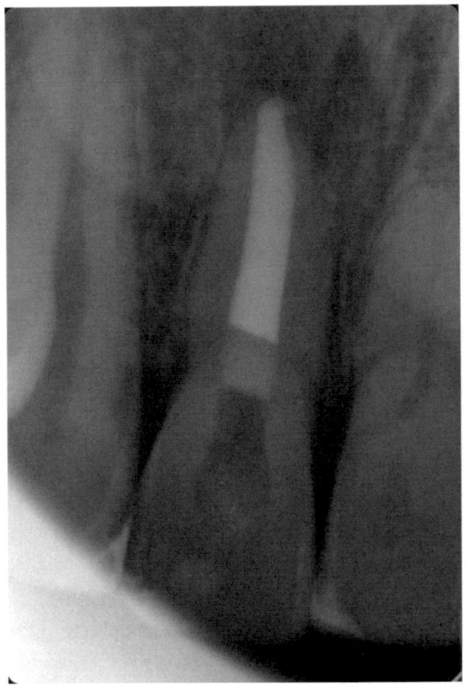

Figure 4.
Open apex and wide foramen apicale, coronal part is filled with internal bleaching aget, as a complication of the treatment, tooth discolaration was observed.

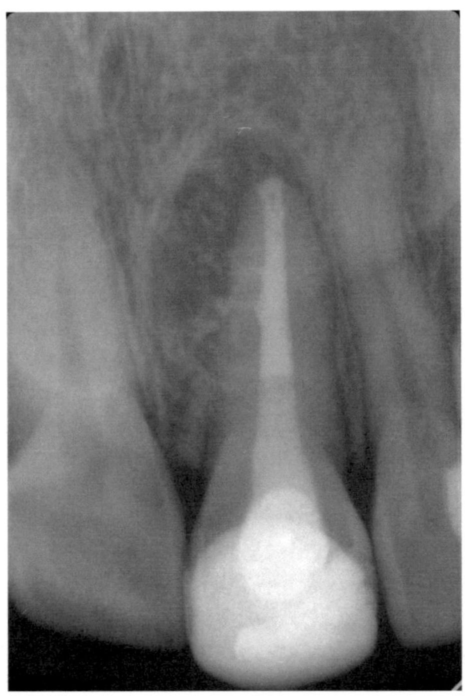

Figure 5.
Open apex and wide root canal system filled with calcium silicate cement.

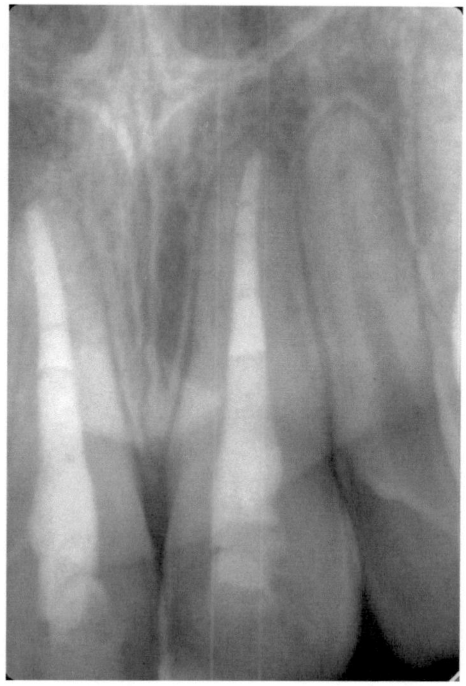

Figure 6.
Apical part of the tooht is filled with calcium silicate material.

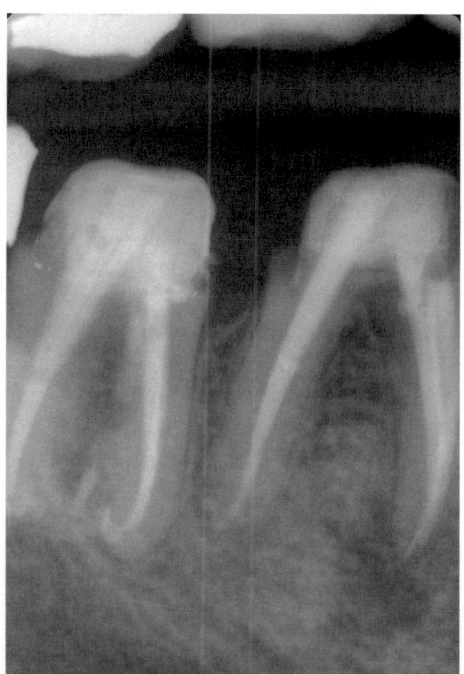

Figure 7.
Perforation repair.

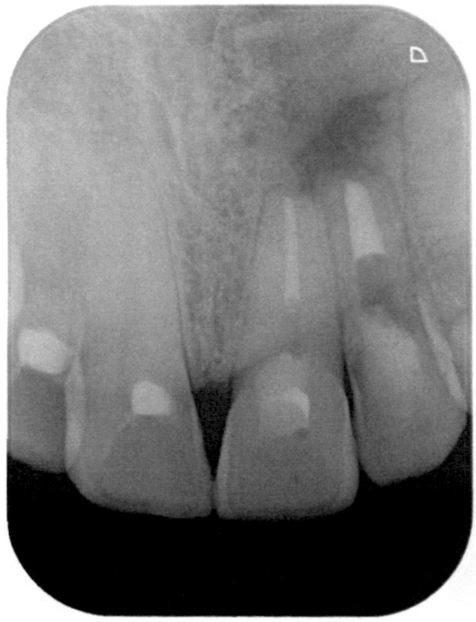

Figure 8.
Calcium silicate cement placed ortogradialy followed by root resection and cystic lession removal facilitate surgical procedure.

2.3.5 Endodontic apical microsurgery

The calcium silicate cements with high biokompatibility, setting in wet conditions, antimicrobioal activity and bonding to dentin present highly indicated materials for endodontic microsurgery. Handling of the material is very important for the very difficult conditions in apical microsurgery. The visibility and the application in limited space are challenges. Radiologic control is neccesery. Fast set materials are advantageous. Putty concistency materials can be placed comfortably for the clinician (**Figure 8**) [39].

2.3.6 External and internal resorption repair

Dental tissue resorption can occur after trauma, orthodontic treatment, chronic inflamatory dissease in pulp or in periodontal tissues. Predentin and precement layer are protecting the tissues before clastical cells. In general, two big groups of resoption exist: internal and external. Internal resorption origins in root canal, and it is the consequence of chronicaly inflamatory changed pulp. Even in condition when part of the pulp is necrotic, the apical part can be inflamatory changed and caused resorption. Root canal treatment is indicated and calcium silicate sealers or calcium silicate cements can be option for root canal obturation. External resorption exists in four forms: 1. external surface resorption, which is transient and requires no treatment; 2. invasive cervical resorption which can be consequent of trauma, intracoronal bleaching, orthodontic, surgical or periodontal treatment. Based on its range the surgical approach, endodontical approach or extraction can be indicated. Pulp tissue is protected by predentin layer. The vitality of the teeth should be protected, but it is sacrificed when this approach is less invasive due to apical extent of resorption. Calcium silicate cements can be used to root canal obturation or resorption cavity filling. This can be done also with composite resin or glass ionomer. 3. Inflamatory root resorption is often consequence of trauma and infected necrosis of the pulp. The bacteria and interleucine induct inflammation which leads to prograssive root resorption. Open apex can be created this way, then the calcium silicate materials in apical portion of the root canal are indicated for obturation as descused before. 4. Replacement resorption is connected with luxation and necrosis of periontal ligaments. The tooth itself is gradualy replaced by bone [40, 41].

2.3.7 Regenerative endodontics

Immature necrotic teeth can be revascularised by intracanal bleeding stabilised by calcium silicate cement. Advantageous is cement with low tendention to discolouration and fast setting. The radiological dentin-like structure is then observed to form in the canal—immitating the root development. From mechanical point of view, it is not clear if the resistence of the tooth is going to get stronger to occlusal forces. The dentin in cervical area will not be formed [33].

There is a histological finding not confirming dentin production. The mineralised tissue does not adhere to dentinal walls [42].

2.3.8 Root canal sealers

Root canal sealers are group of cements used with guttapercha to fill the root canal space, where are irregularieties between gutapercha and root canal walls block further bacterial growth and isolate remaining bacterias from nutrients. The hermetic seal is desired. Sealer can fill lateral canals, voids and apical delta. Root canal sealers should have optimal mixing and setting time (to allow proper obturation), radioopacity, reology and should not discolourate teeth. It is desirable to be antimicrobiologicly active and not to cause inflamatory respons or cytotoxicyty. Bind to the dentinal walls of the canal and tacky consistency. It is required to contain fine powder. Insolubility is desired as well as posibility of removal during retreatment. Preferably biocative in nature. Ideal root canal sealer does not exist. MTA-based sealers fullfil many of desired characteristics (**Figure 9**) [43, 44].

3. Cement-based material in prosthodontics

3.1 Provisional cements

Temporary cementation is suitable for temporary restorations, which will soon be replaced by definitive ones. Temporary restoration has the following advantages: 1. protection of the abutment against mechanical, thermal and chemical damage; 2. gingiva shaping or definition of ideal gingival zenith (using biologically oriented preparation technique); 3. protection of finish line against damage; 4. preservation of spatial conditions—prevents the inclination of adjacent teeth, prevents the tooth from entering supraocclusion; 5. aesthetically and functionally replaces lost tissues. In case the final restoration is cemented adhesively, the temporary cement must not contain eugenol. Temporary cements are usually oil-based. Modern temporary cements are oil-based zinc oxide without eugenol. Compared with final cements, they have weaker physical properties and form a thicker layer of material. Temporary cement is also suitable for office cementation of suprastructures on implants. It will allow us greater

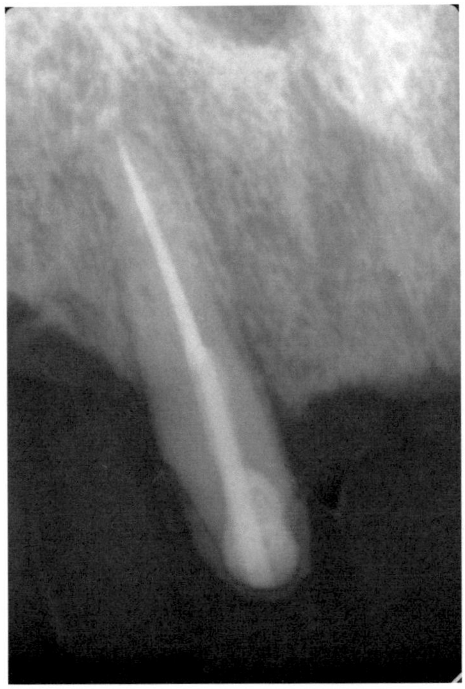

Figure 9.
Root canal is filled with calcium silicate cement based sealer, composite cement is used for core build-up and provisional restoration is cemented with provisional oil-based cement.

repairability of protective work compared with final cements. Cemented implant work is rarely used, excess cement can cause periimplantitis. Screwed restoration is more suitable [45, 46].

The requirements for temporary cements are: biocompatibility, easy removal from the tooth surface, easy removal from the restoration surface, easy material preparation, easy removal of excess, clinically acceptable setting time, temporary cement must allow easy removal of the temporary restoration but ensure good retention. It must not react with the final cement [47].

Temporary cementation of definitive restoration is high risk! Abutment teeth can be damaged when removing the final prosthesis. The author does not recommend the use of temporary cements for big prosthetic reconstructions. It is more appropriate to extend the time with temporary prosthetic work than to temporarily cement the final work. In addition, temporary work can be easily modified. Temporary cement must be thoroughly removed before final cementation, residues of temporary cement preclude the achievement of maximum efficiency of the final cement. In a special case—that is, during reconstruction with aesthetic veneers—when preparation is non-retentive, a point-adhesive-prepared surface and composite cement can be used to cement the temporary veneers. The size of the adhesively prepared surface in clinical practice is approximately 1 mm × 1 mm. There is also a possibility to use mock up [48].

3.1.1 Definitive cements

Definitive cements can be divided in two groups: water-based and adhesive cements (composite resin based) [1].

In addition to cement, the retention of the restoration is also ensured by the preparation itself and the bevel angle of the preparation. At the same bevel angle, the adhesive cements provided twice the retention force than conventional glass ionomer cement or zinc oxide phosphate cement. At a 24 degree preparation angle, the retention force was 20% higher using resin cements than at a 6 degree preparation angle using conventional cements. This demonstrates the higher retentive properties of adhesive cements [49].

We generally require low viscosity from cements so that they form the smallest possible layer [50].

The preparation of surfaces, which are connected with cements, also varies.

3.1.2 Water-based cements (acidobasic cements)

When setting, they go through acido-basic reaction. This may explain the transient increased pulpal sensitivity after cementation [51, 52].

They are not as susceptible to the presence of water as hydrophobic resin cements. It is clinically appropriate to provide a relative dry field. We use it advantageously for deep preparations, where the moisture control is problematic.

The main represenatatives of watter-based cements are: zinc oxide phosphate, glass ionomers and resin-modified glass ionomers.

3.1.2.1 Zinc oxide phosphate

Zinc oxide phosphate does not bond to teeth structure. What is advantageous is easy removal of excesses. High early strength makes it suitable for cementing metal cast posts. The material is characterised by high compressive strength and low tensile strength. There is very low Ph during solidification. It is well soluble at first. Clinican should be aware of contamination with saliva [53].

3.1.2.2 Glass ionomers

Although they are well-known as glass ionomers, the correct name for them according to the International Organisation for Standardisation is glass polyalekenoate cements. It is a mixture of weak acids and with glasses, which is chemically based and except of setting reaction, it works as a filler. Glass ionomers are characterised by binding to hard dental tissues. This bond is mediated on a mineral basis and is slightly higher to the enamel (2.6–9.6 MPa) than to the dentin (1.1–4.1 MPa), but the failure in most studies was cohesive. The tensile strength of glass ionomer cements is relatively low. It solidifies in minutes, can be mixed by hand or in a blender and be applied from a capsule. The acids used in glass ionomer cements are polyalkenoic acids, either as a homopolymer of acrylic acid or a copolymer of acrylic acid and maleic acid. The contained glass is alumino-silicate with floride and phosphate. Without aluminimum, the glass only from SiO_4 would not be reactive with acids. Al caries 3+ charge and when implemented into tetraedrical geometry of silicon—the glass become base [1, 53].

Solidification takes place in several stages. It usually takes 2–3 min before they lost flow (sometimes up to 6 min). Early solidification of the material leads to an improper fit of prosthetic work or to the impossibility of applying this material to the cavity. The first phase of solidification is just the reaction of weak acids and glass, ions such as Ca, Na move. Subsequently, cross-linking occurs, for which Al ions are responsible. It takes about a day. They are initially sensitive to water, premature drying out or early matrix removal can lead to rough surfaces [53].

Water is the medium in which the reaction takes place and in which the acids are dissolved. Excessive drying and cracking of the surface can occur during solidification—this is prevented by varnishes from resin.

Glass ionomer cements can be used as a fixing material, but also as a restorative material as mentioned before. The properties it should achieve are defined by the iso standard. In both cases, they are different. The minimum compressive strength for luting material is 70 MPa [53].

Glass ionomers are bonding chemically to hydroxipapatit. As advantage can be seen that they are releasing fluorides, this property decreases with its maturation [53].

Disadvatage is mainly in restorative dentistry as glass ionomer fillings are not so mechanically resistant to function as a long-lasting definitive filling. Fixation cements are radiopaque to check for excess removal.

Traditionally, there are three types: 1. fixation, 2. restorative, 3. lining and base [53].

For fixation the dentin can be slightly wet, but with no watter or saliva pooling on the surface.

Cementation in prosthetics by glass ionomer is indicated especially for these materials: metal and metal-ceramic crowns, zirconium ceramics . Otis indicated for this prosthetic works: crowns, fixed partial denture and root inlay. The onlay or adhesive bridges cannot be cemented by water-based cements as the preparation itself is not retentive. The indication does not differ for other watter-based cements.

3.1.2.3 Resin-modified glass ionomers

The improvement of mechanical properties and handling of glass ionomer materials was created by their modification with a resin matrix. Cements are more mechanically resistant and are not so prone to early water contamination in contrast to conventional glass ionomer cements. The solidification of the resin matrix can be ensured chemically. However, the content of HEMA compromises their biocompatibility [54].

3.1.2.4 Adhesive cements

For adhesive fixation, we use composite cements (based on methacrylate resin with different contents of monomer, fillers and silane) light-curing or dual-curing. Dual-curing has the advantage of curing even under a prosthesis that is opaque or too large for light to penetrate, as the light begins polymerisation. Low translucency lithium disilicate discs showed improper degree of conversion under 1.2 mm thick material for light-curing cements [55, 56] (**Figure 10**).

Light-curing materials show less tendency to colour change in time than dual-curing materials. They are therefore more suitable when there is high aesthetic demand [57].

Adhesive cementation of the etchable ceramic leads to a significant increase in the flexural strength of the ceramic. However, this is sensitive to water contamination. For long-term success, it is important to maintain absolute dry field during cementation secured by rubberdam [58] (**Figures 11** and **12**).

If it is possible to fix the ceramic work adhesively, it is recommended, because it also improves the mechanical properties of the ceramic itself, especially in bending.

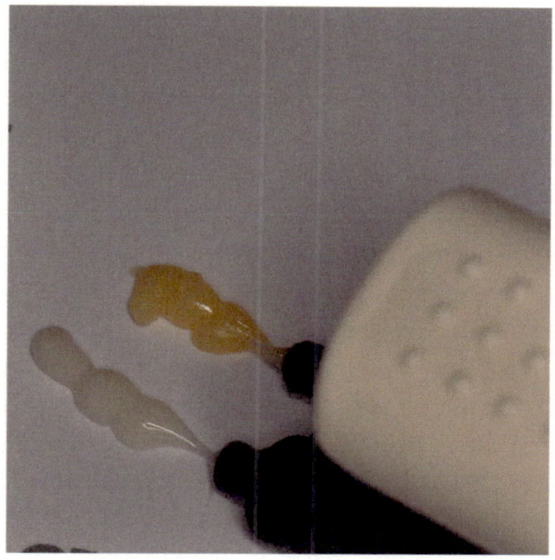

Figure 10.
Example of adhesive cement before mixing.

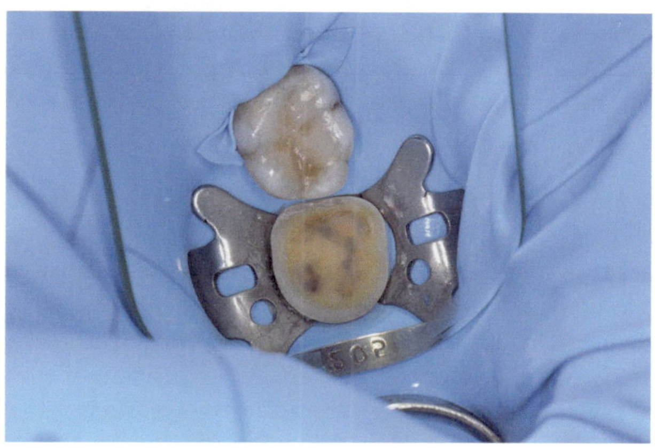

Figure 11.
Lower second molar before adhesive cementation. Dry field is maintained by rubberdam.

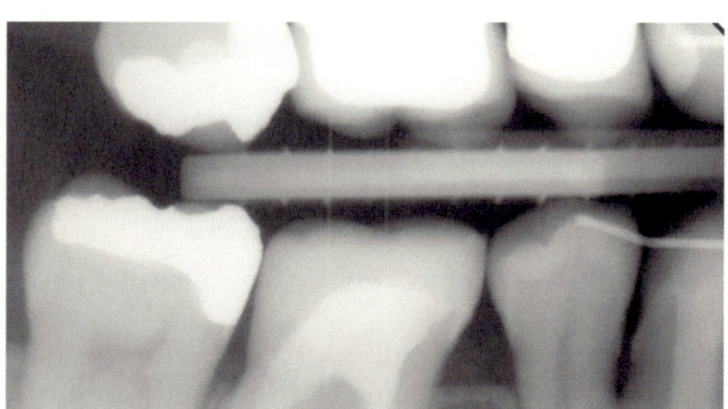

Figure 12.
X ray showing the fit and the lack of cement excesses after cementation of lower left first molar.

We use them if the preparation of the tooth pillar itself does not provide good retention.

For example, veneer preparations do not provide retention. It is therefore advisable to use adhesion to the enamel. Seventy percent of the enamel should be considered as a recommended.

Materials with a high glass content provide the possibility of high aesthetics such as transparent incisions. However, the amorphous ceramic glass and the low content of crystals (responsible for the mechanical resistence) require a very strong adhesive fixation to ensure the resistance of the restoration.

It is easier to achieve higher adhesion to enamel than to dentin. Adhesive preparation of dentin for indirect replacement should be performed immediately, collagen fibres are not exposed to saliva, bacteria are prevented from accessing the pulp. IDS or resin coating thus takes over some of the functions of the temporary restoration [59].

American Dental Association (ADA) specification no. 8 and ADA/ANSI specification No. 96 defined ideal thickness of the material of the cementation as 25 um or less. This is important for precise setting of the restoration and also for dental technician to fabricate the restoration which is satisfying even after adding material of the cementation. Depending on the material of the restoration, the luting cement colour and opacity can slightly influence the perceived colour. In the case of severly discoloured abutment teeth, the desired colour change is probably not achievable by 25 um of cement material. It is reasonable to change the colour in the prosthetic restoration [60–62].

3.1.2.5 Adhesion to hard dental tissues

Adhesion to natural hard dental tissues involves two basic approaches: total etch (also known as etch and rinse) and self-etch adhesive. Composite cements used for total etch approach require etching by 37% phosporoc acid (enamel 30 s and dentin 10 s) and rinsing. After this procedure, apllication of the adhesive (one or two botles) is possible. Although the adhesive system is self-etch, cements connecting acid monomers are self-etch, self-adhesive. By this the application of the adhesive system is eliminated. Self-etch self-adhesive cements contain acid monomer, which is etching and priming tooth structure. This leads to reducing steps in cementation procedure. Etch and rinse systems lead to creating more retentive pattern in enamel. Total etch approach is superior when the retention is mostly to enamel. Effectivness of self-etch self-adhesive cements can be improved by selective etching [A] (**Figures 13** and **14**).

Bonding to dentin is complicated by higher organic and watter contain. The bonding strength decreases over time. The bonding strength of self-etch adhesives is more stable in time in dentin. Self-etch approach offers this advantage when bonding to the large areas of dentin [63–65].

Self-etch self-adhesive cements bond suffciently to dentin, but preetching of the enamel is advisable [66, 67].

As was shown in a recent study [67], if only IDS is performer, than the sandblasting exposes the dentin and reetching and pretreating dentin are advisable. Another option is to reinforce IDS with flowable composite, thus creating a resin coating. Then the adhesive procedure is done on composite surface [B].

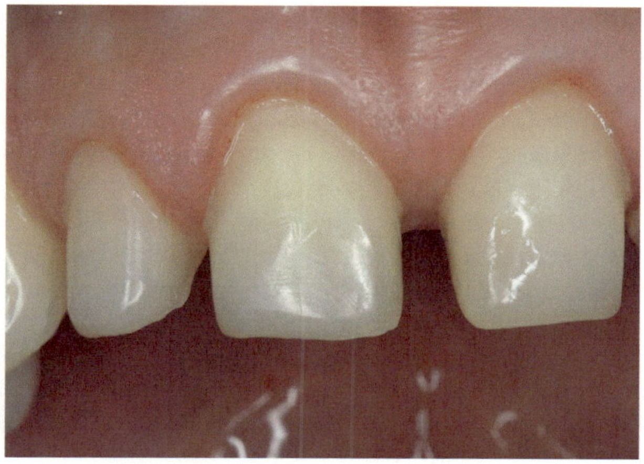

Figure 13.
Veneer preparation—note the preserved enamel.

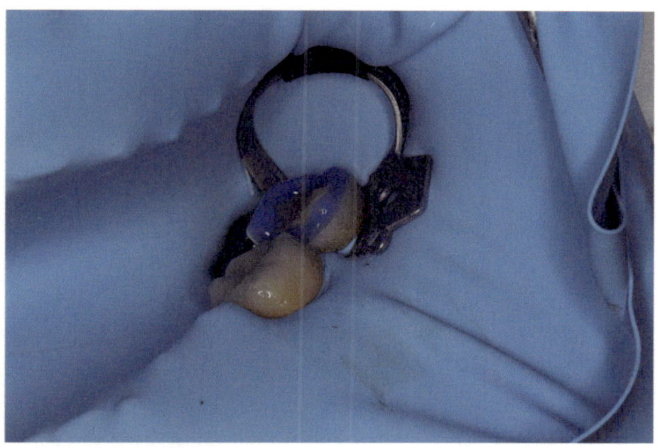

Figure 14.
Selective etching before cementation.

3.1.2.6 Bonding to dental composite

Bonding to dental composite brings several challenges such as no presence of free monomer in the composite (during the direct fabrication of the composite filling the layering technig utilise the inhibition layer on the surface of the composite by which next layer is connected). The highest bonding strength to industrial composite was achieved by mechanical and chemical conditioning. Mechanical sand blasting with alumina—air abrasion. Chemical pretreating by etching, application of universal primer and silane. This was suggested as a best strategy by meta-analysis from 2019. Other strategies reducing steps were also suggested. However, pretreatment of the surface of composite is indicated [68].

The sandblasting, rinsing with water and air drying are also efficient ways of preparing composite surface to applying adhesive [69].

3.1.2.7 Bonding to ceramic with glass contain

Ceramic with glass offers high estetic solution. However, it is suspectible to fracture. The lithium disilicate ceramic (representative of resistant but esthetic ceramic) can with thickness 0.3 mm offer a flexural strength of 400 MPa (**Figure 15**).

Ceramic restoration is etched by HF 4–9%. The time of etching depends on glass. For lithium disilate ceramic, the time of etching is 20 s. The ceramic restoration is rinsed with water spray and can be placed into ultrasonic cleaner. Salt hexaflouro-silicate is released. The crystals are exposed and micropores for resin cement are accesible. Silane (SiH4) reacts with OH group (inorganic part) and with methacrylate group (organic part), thus connecting restoration and cement. Feldspatic ceramic with high content of glass should be etched in 60 s, but is also vulnerable to over-etching. Ceramic with high content of Al_2O_3 even after etching is not sufficiently prepared. For such a ceramic sandblasting can create microporosites. This requires very resistant ceramic and ceramic with higher content of glass could be broken (**Figures 16–18**) [70–72].

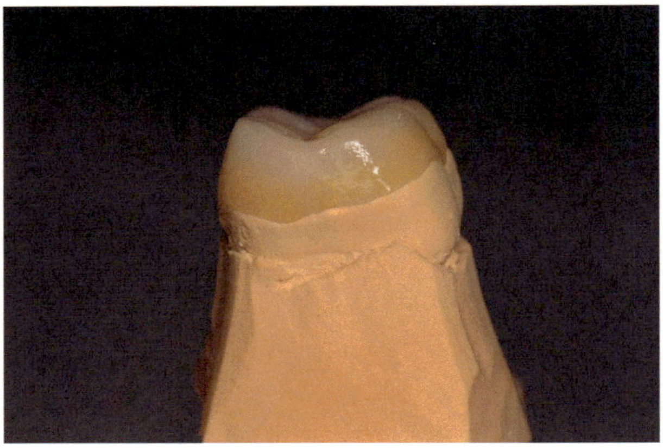

Figure 15.
Indirect prosthetic restoration before cementation.

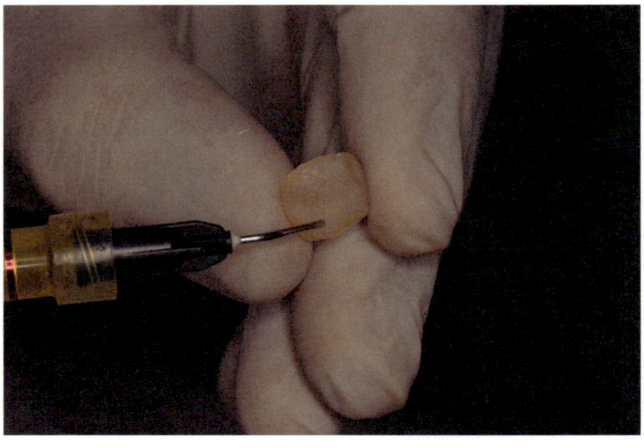

Figure 16.
Surface of ceramic with low glass contain, application of 4% HF.

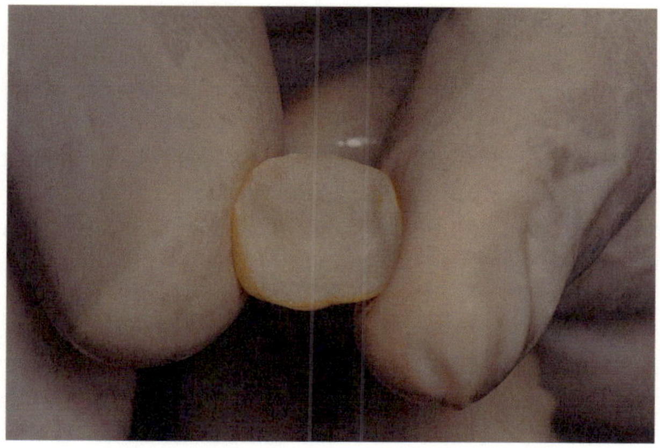

Figure 17.
Surface of ceramic with low glass contain after etching using 4% HF.

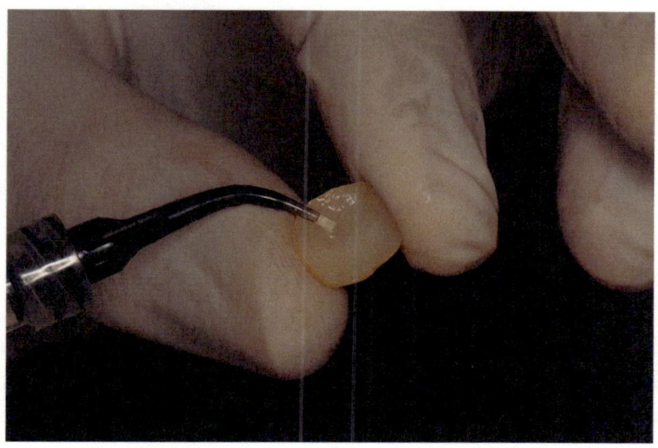

Figure 18.
Silane applied on indirect restoration with low glass contain.

3.1.2.8 Bonding to the polycrystalic ceramic

The flexural strength of ZrO2-based ceramics is up to two times greater than lithium disilicate ceramic. Cementation in relative dry field on glassionomer cement is clinicaly easy. However, some clinical conditions ask for adhesive cementation to ensure higher bonding strength and hopefully longevity of the restoration. For example, parafunctional habbit requires high resistant restoration and the length of clinical crown can be compromised, this is a combination of risk factors, which is dificult to solve without polycrystalic ceramic and without adhesive cementation. Polycrystalic zirconia ceramic can refer to differently marked ceramics such as 3Y, 4Y and 5Y. This refers to Yitria content, with the higher content of Yitria, the higher translucency can be achieved. Adversely, with higher Yitria content, the less flexural strength of the ceramic. The damage of the ceramic through sandblasting is neglectable. The same protocol can be applied on 3Y, 4Y and 5Y zirconia. Air abrasion followed by application of special primer with monomer 10-methacryloyloxydecyl dihydrogen phosphate (MDP) or composite resin cements containing MDP can lead to the long-term bond [73–75].

4. Conclusion

Calcium silicate cements and adhesive cementation of different materials offer many advantages, which changed dental medicine. Less invasive approach is now possible due to development in the field of cements. Nevertheless, these materials have their own disadvantages and challenges. Clinicans should keep in mind both and not to believe in marvellous material, but to know their own materials, their requirements and still follow basic principles.

Acknowledgements

Authors appreciate support from Klinika zubního lékařství LF a FN which creates incredible working condition for both scientific and dental work.

The chapter was supported by funds IGA LF 2022 043 and IGA LF 2022 021.

Conflict of interest

The authors declare no conflict of interest.

Thanks

To my parents for continuous support and encouragement during work as a scientist and doctor.

To Abanoub Riad, PhD. for humble, hard-working scientific work, which inspire me.

To Dr. Pavel Holík, PhD., Dr. Roman Moštěk, Dr. Matouš Kašpar and Dr. Matěj Rosa for their friendship and for their X-rays which they provided to me.

To the women I love.

Author details

Ján Staněk,* Basel Elia Azar and Tomáš Fichtel
Faculty of Medicine and Dentistry, Institute of Dentistry and Oral Sciences, Palacký University, Olomouc, The Czech Republic

*Address all correspondence to: stanekjano@gmail.com

References

[1] Sanz JL, Guerrero-Gironés J, Pecci-Lloret MP, Pecci-Lloret MR, Melo M. Biological interactions between calcium silicate-based endodontic biomaterials and periodontal ligament stem cells: A systematic review of in vitro studies. International Endodontic Journal. 2021;**54**(11):2025-2043

[2] Chen Y, Chen X, Zhang Y, Zhou F, Deng J, Zou J, et al. Materials for pulpotomy in immature permanent teeth: A systematic review and meta-analysis. BMC Oral Health. 2019;**19**(1):227

[3] Li Y, Sui B, Dahl C, Bergeron B, Shipman P, Niu L, et al. Pulpotomy for carious pulp exposures in permanent teeth: A systematic review and meta-analysis. Journal of Dentistry. 2019;**84**:1-8

[4] Dos Santos RA, de Lima EA, Mendonça LS, de Oliveira JE, Rizuto AV, de Araújo Silva Tavares ÁF, et al. Can universal adhesive systems bond to zirconia? Journal of Esthetic and Restorative Dentistry. 2019;**31**(6):589-594

[5] Zhu YJ, Guo XX, Sham TK. Calcium silicate-based drug delivery systems. Expert Opinion on Drug Delivery. 2017;**14**(2):215-228

[6] Koko M, Takagaki T, Abd El-Sattar NEA, Tagami J, Abdou A. MDP Salts: A New Bonding Strategy for Zirconia. Journal of Dental Research. 2022 Jul;**101**(7):769-776

[7] Shu C, Zheng X, Wang Y, Xu Y, Zhang D, Deng S. Captopril inhibits matrix metalloproteinase activity and improves dentin bonding durability. Clinical Oral Investigations. 2022;**26**(3):3213-3225

[8] Retana-Lobo C, Guerreiro-Tanomaru JM, Tanomaru-Filho M, Mendes de Souza BD, Reyes-Carmona J. Sodium hypochlorite and chlorhexidine downregulate MMP expression on radicular dentin. Medical Principles and Practice. 2021;**30**(5):470-476

[9] Instruments and Equipment. "ANSI/ADA specification no. 66 for dental glass ionomer cements. Council on Dental Materials, Instruments, and Equipment." Journal of the American Dental Association. 1989;**119**(1):205

[10] Bobotis HG, Anderson RW, Pashley DH, Pantera EA Jr. A microleakage study of temporary restorative materials used in endodontics. Journal of Endodontia. 1989;**15**(12):569-572

[11] Weiner RS. Liners, bases, and cements: A solid foundation. General Dentistry. 2002;**50**(5):442-446

[12] Kim D, Kim E. Antimicrobial effect of calcium hydroxide as an intracanal medicament in root canal treatment: A literature review—Part I. In vitro studies. Restorative Dentistry and Endodontics. 2014;**39**(4):241-252

[13] National Center for Biotechnology Information. "PubChem Patent Summary for US-5769638-A, Tooth filling material and method of use" PubChem. Available from: https://pubchem.ncbi.nlm.nih.gov/patent/US-5769638-A [Accessed 6 Aug 2022]

[14] Altan H, Tosun G. The setting mechanism of mineral trioxide aggregate. Journal of Istanbul University and Faculty of Dentistry. 2016;**50**(1):65-72

[15] Shokouhinejad N, Nekoofar MH, Razmi H, Sajadi S, Davies TE, Saghiri MA, et al. Bioactivity of EndoSequence root repair material and bioaggregate. International Endodontic Journal. 2012;**45**(12):1127-1134

[16] Sarkar NK, Caicedo R, Ritwik P, Moiseyeva R, Kawashima I. Physicochemical basis of the biologic properties of mineral trioxide aggregate. Journal of Endodontia. 2005;**31**(2):97-100

[17] Parirokh M, Torabinejad M. Mineral trioxide aggregate: A comprehensive literature review—Part I: Chemical, physical, and antibacterial properties. Journal of Endodontia. 2010;**36**(1):16-27

[18] Torabinejad M, Hong CU, McDonald F, Pitt Ford TR. Physical and chemical properties of a new root-end filling material. Journal of Endodontia. 1995;**21**(7):349-353

[19] Fridland M, Rosado R. MTA solubility: A long term study. Journal of Endodontia. 2005;**31**(5):376-379

[20] Linsuwanont P. MTA apexification combined with conventional root canal retreatment. Australian Endodontic Journal. 2003;**29**(1):45-49

[21] Marciano MA, Duarte MA, Camilleri J. Dental discoloration caused by bismuth oxide in MTA in the presence of sodium hypochlorite. Clinical Oral Investigations. 2015;**19**(9):2201-2209

[22] Duarte MAH, Alves de Aguiar K, Zeferino MA, Vivan RR, Ordinola-Zapata R, Tanomaru-Filho M, et al. Evaluation of the propylene glycol association on some physical and chemical properties of mineral trioxide aggregate. International Endodontic Journal. 2012;**45**(6):565-570

[23] Abdullah D, Pitt Ford TR, Papaioannou S, Nicholson J, McDonald F. An evaluation of accelerated Portland cement as a restorative material. Biomaterials. 2002;**23**(19):4001-4010

[24] Lee SJ, Monsef M, Torabinejad M. Sealing ability of a mineral trioxide aggregate for repair of lateral root perforations. Journal of Endodontia. 1993;**19**(11):541-544

[25] Basturk FB, Nekoofar MH, Gunday M, Dummer PMH. Effect of varying water-to-powder ratios and ultrasonic placement on the compressive strength of mineral trioxide aggregate. Journal of Endodontia. 2015;**41**(4):531-534

[26] Shahravan A, Jalali SP, Torabi M, Haghdoost AA, Gorjestani H. A histological study of pulp reaction to various water/powder ratios of white mineral trioxide aggregate as pulp-capping material in human teeth: A double-blinded, randomized controlled trial. International Endodontic Journal. 2011;**44**(11):1029-1033

[27] Shahi S, Rahimi S, Yavari HR, Samiei M, Janani M, Bahari M, et al. Effects of various mixing techniques on push-out bond strengths of white mineral trioxide aggregate. Journal of Endodontia. 2012;**38**(4):501-504

[28] Basturk FB, Nekoofar MH, Günday M, Dummer PM. The effect of various mixing and placement techniques on the compressive strength of mineral trioxide aggregate. Journal of Endodontia. 2013;**39**(1):111-114

[29] Hilton TJ, Ferracane JL, Mancl L, Northwest Practice-based Research Collaborative in Evidence-based Dentistry (NWP). Comparison of CaOH with MTA for direct pulp capping: A PBRN randomized clinical trial. Journal of Dental Research. 2013;**92**(7):16S-22S

[30] Matsuo T, Nakanishi T, Shimizu H, Ebisu S. A clinical study of direct pulp capping applied to carious-exposed

pulps. Journal of Endodontia. 1996;**22**(10):551-556

[31] Hafez AA, Cox CF, Tarim B, Otsuki M, Akimoto N. An in vivo evaluation of hemorrhage control using sodium hypochlorite and direct capping with a one- or two-component adhesive system in exposed nonhuman primate pulps. Quintessence International. 2002;**33**(4):261-272

[32] American Association of Endodontists. Glossary of endodontic terms. 10th ed. Chicago: American Association of Endodontists; 2020. Available from: https://www.aae.org/specialty/clinical-resources/glossary-endodontic-terms/ [Accessed 14 Sep 2021]

[33] Ricucci D, Siqueira JF Jr, Li Y, Tay FR. Vital pulp therapy: Histopathology and histobacteriology-based guidelines to treat teeth with deep caries and pulp exposure. Journal of Dentistry. 2019;**86**:41-52

[34] Storm B, Eichmiller FC, Tordik PA, Goodell GG. Setting expansion of gray and white mineral trioxide aggregate and Portland cement. Journal of Endodontia. 2008;**34**(1):80-82

[35] Juarez Broon N, Bramante CM, de Assis GF, Bortoluzzi EA, Bernardineli N, de Moraes IG, et al. Healing of root perforations treated with Mineral Trioxide Aggregate (MTA) and Portland cement. Journal of Applied Oral Science. 2006;**14**(5):305-311

[36] Holland R, Filho JA, de Souza V, Nery MJ, Bernabe PF, Junior ED. Mineral trioxide aggregate repair of lateral root perforations. Journal of Endodontics. 2001;**27**(4):281-284

[37] Main C, Mirzayan N, Shabahang S, Torabinejad M. Repair of root perforations using mineral trioxide aggregate: A long-term study. Journal of Endodontia. 2004;**30**(2):80-83

[38] von Arx T. Apical surgery: A review of current techniques and outcome. Saudi Dental Journal. 2011;**23**(1):9-15

[39] Ne RF, Witherspoon DE, Gutmann JL. Tooth resorption. Quintessence International. 1999;**30**(1):9-25

[40] Heithersay GS. Management of tooth resorption. Australian Dental Journal. 2007;**52**(1 Suppl):S105-S121

[41] Banchs F, Trope M. Revascularization of immature permanent teeth with apical periodontitis: New treatment protocol? Journal of Endodontia. 2004;**30**(4):196-200

[42] Camilleri J, Gandolfi MG, Siboni F, Prati C. Dynamic sealing ability of MTA root canal sealer. International Endodontic Journal. 2011;**44**(1):9-20. DOI: 10.1111/j.1365-2591.2010.01774.x PMID: 20646079

[43] Morgental RD, Vier-Pelisser FV, Oliveira SD, Antunes FC, Cogo DM, Kopper PM. Antibacterial activity of two MTA-based root canal sealers. International Endodontic Journal. 2011;**44**(12):1128-1133

[44] Loi I, Di Felice A. Biologically oriented preparation technique (BOPT): A new approach for prosthetic restoration of periodontically healthy teeth. The European Journal of Esthetic Dentistry. 2013;**8**(1):10-23

[45] Vahidi F. The provisional restoration. Dental Clinics of North America. 1987;**31**(3):363-381

[46] Abrams SH. Current concepts in temporary cement. Oral Health. 1995;**85**(4):19

[47] Vinod Kumar G, Soorya Poduval T, Reddy B, Shesha

Reddy P. A study on provisional cements, cementation techniques, and their effects on bonding of porcelain laminate veneers. Journal of Indian Prosthodontics Society. 2014;**14**(1):42-49. DOI: 10.1007/s13191-012-0219-5

[48] Zidan O, Ferguson GC. The retention of complete crowns prepared with three different tapers and luted with four different cements. The Journal of Prosthetic Dentistry. 2003;**89**(6):565-571

[49] American Dental Association (ADA) specification no. 96

[50] Akpata ES, Sadiq W. Post-operative sensitivity in glass-ionomer versus adhesive resin-lined posterior composites. American Journal of Dentistry. 2001;**14**(1):34-38

[51] Chandrasekhar V. Post cementation sensitivity evaluation of glass Ionomer, zinc phosphate and resin modified glass Ionomer luting cements under class II inlays: An in vivo comparative study. Journal of Conservation Dental. 2010;**13**(1):23-27

[52] Glockmann E, Glockmann I, Hörenz D, Lange G, Reichardt R, Schwarzburg G. Untersuchungen an Zinkoxid-Phosphate-Zementen hinsichtlich Anwendbarkeit des Firmat [Zinc oxide phosphate cement studies on the usefulness of Firmat]. Stomatol DDR. 1976;**26**(1):7-15

[53] Sidhu SK, Nicholson JW. A review of glass-ionomer cements for clinical dentistry. Journal of Functional Biomaterials. 2016;**7**(3):16

[54] Wilson AD. Resin-modified glass-ionomer cements. The International Journal of Prosthodontics. 1990;**3**(5):425-429

[55] Cho SH, Lopez A, Berzins DW, Prasad S, Ahn KW. Effect of different thicknesses of pressable ceramic veneers on polymerization of light-cured and dual-cured resin cements. Journal of Contemporary Dental Practice. 2015;**16**(5):347-352

[56] Pegoraro TA, da Silva NR, Carvalho RM. Cements for use in esthetic dentistry. Dental Clinics of North America. 2007;**51**(2):453-471

[57] Pissaia JF, Guanaes BKA, Kintopp CCA, Correr GM, da Cunha LF, Gonzaga CC. Color stability of ceramic veneers as a function of resin cement curing mode and shade: 3-year follow-up. PLoS One. 2019;**14**(7):e0219183

[58] Addison O, Marquis PM, Fleming GJ. Adhesive luting of all-ceramic restorations--the impact of cementation variables and short-term water storage on the strength of a feldspathic dental ceramic. The Journal of Adhesive Dentistry. 2008;**10**(4): 285-293. PMID: 18792699

[59] Magne P. Immediate dentin sealing: A fundamental procedure for indirect bonded restorations. Journal of Esthetic and Restorative Dentistry. 2005;**17**(3):144-154

[60] American Association of Endodontists. Glossary of Endodontic Terms, 10th edition. Available from: www.aae.org/glossary [Accessed 22 Nov 2020]

[61] EL-Bieh AY, Katamesh H, EL-Agrodi M. The effect of resin cement on masking the color of stained teeth. Cairo Dental Journal. 2012;**28**(3):1279-1287

[62] Vafaee F, Heidari B, Khoshhal M, Hooshyarfard A, Izadi M, Shahbazi A, et al. Effect of Resin Cement Color on the Final Color of Lithium Disilicate All-Ceramic Restorations. Journal of

Dentistry (Tehran). 2018 May;**15**(3): 143-150. PMID: 30090114; PMCID: PMC6079184

[63] Koshiro K, Inoue S, Tanaka T, Koase K, Fujita M, Hashimoto M, et al. In vivo degradation of resin-dentin bonds produced by a self-etch vs. a total-etch adhesive system. European Journal of Oral Sciences. 2004;**112**(4):368-375

[64] Hammal M, Chlup Z, Ingr T, Staněk J, Mounajjed R. Effectiveness of dentin pre-treatment on bond strength of two self-adhesive resin cements compared to an etch-and-rinse system: An in vitro study. PeerJ. 2021;**9**:e11736

[65] Ozer F, Blatz MB. Self-etch and etch-and-rinse adhesive systems in clinical dentistry. Compendium of Continuing Education in Dentistry. 2013;**34**(1):12-16

[66] Rosa WL, Piva E, Silva AF. Bond strength of universal adhesives: A systematic review and meta-analysis. Journal of Dentistry. 2015;**43**(7):765-776

[67] Kovalsky BT, Voborna I, Ingr T, Morozova Y, Misova E, Hepova M. Immediate dentin sealing: Effect of sandblasting on the layer thickness. Bratislava Medical Journal. 2022;**123**(2): 87-91. DOI: 10.4149/BLL_2022_015

[68] Yu H, Özcan M, Yoshida K, Cheng H, Sawase T. Bonding to industrial indirect composite blocks: A systematic review and meta-analysis. Dental Materials. 2020;**36**(1):119-134

[69] D'Arcangelo C, De Angelis F, D'Amario M, Zazzeroni S, Ciampoli C, Caputi S. The influence of luting systems on the microtensile bond strength of dentin to indirect resin-based composite and ceramic restorations. Operative Dentistry. 2009;**34**(3):328-336

[70] Santos GC Jr, Santos MJ, Rizkalla AS. Adhesive cementation of etchable ceramic esthetic restorations. Journal of the Canadian Dental Association. 2009;**75**(5):379-384

[71] Kamada K, Yoshida K, Atsuta M. Effect of ceramic surface treatments on the bond of four resin luting agents to a ceramic material. The Journal of Prosthetic Dentistry. 1998;**79**(5):508-513

[72] Roulet JF, Söderholm KJ, Longmate J. Effects of treatment and storage conditions on ceramic/composite bond strength. Journal of Dental Research. 1995;**74**(1):381-387

[73] Janda R, Roulet JF, Wulf M, Tiller HJ. A new adhesive technology for allceramics. Dental Materials. 2003;**19**(6):567-573

[74] Qualtrough AJ, Piddock V. Dental ceramics: What's new? Dental Update. 2002;**29**(1):25-33

[75] Alammar A, Blatz MB. The resin bond to high-translucent zirconia-A systematic review. Journal of Esthetic and Restorative Dentistry. 2022;**34**(1):117-135

Chapter 3

Concrete from Alternative and Waste Materials

Emmanuel Ndububa

Abstract

The manuscript discusses alternative and cheaper concrete types developed through research and practice that attract worldwide attention, especially those that contain components, such as waste materials from industry, agriculture, mining, and domestic. These concretes include post-concrete and geopolymer concrete. Others are synthetic concrete, pozzolana concrete, fiber concrete, and mortar. This alternative concrete has been found to possess improved mechanical strength and durability, especially higher tensile, flexural, shockproof, hardness, crack relief, and resistance to acid and sulfate attacks. The choice of alternative constituent materials has become a potential waste material that can completely or partially replace these mixture components. The production of alternative concrete is essential in the reduction of costs, carbon dioxide emissions, and for environmental sustainability. The purpose of this manuscript is to present to stakeholders in the construction industry some alternative concretes that are economical and environmentally friendly but yet compete with conventional concretes in strength and surpass durability in most cases.

Keywords: alternative concrete, waste, geopolymer, cement, fiber reinforced concrete, environmental sustainability

1. Introduction

Concrete is a brittle composite material that is good in compression but poor in tension used for construction purposes and consists of hard aggregates that are bonded together by cement and water. Most normal or conventional concrete is produced from ordinary portland cement (OPC). The aggregates are usually sand, referred to as *fine aggregates,* and gravels referred to as *coarse aggregates*. The bonding agent is a slurry of *cement* with *water* (referred to as *cement paste*) which reacts together in a process called *hydration*. The concrete sets and hardens over time due to *curing*.

As an important construction material, concrete is in high demand for use in the built environment to erect buildings and construct civil engineering structures like highways, railways, bridges, harbors, and dams. Other structural applications include its use in the manufacture of concrete kerbs, pipes, and open channel drainage system elements.

Concrete is considered the most widely used substance on earth after water [1]. It is considered the base of modern built environment development by providing roofing systems, protection from natural disasters, and providing structures for human endeavors. However, the constituent materials that make up the concrete composite are expensive. The cost of cement is not only high but its production also contributes immensely to carbon dioxide (CO_2) emissions leading to negative climate change. In many parts of the world, fine and coarse aggregates are not only expensive but are scarce commodities. The option for alternative component materials from possibly waste materials that can either fully or partially replace these mixture components to produce alternative concrete has become imperative.

Alternative concretes from cheaper, lighter, and less dangerous component materials are being researched and used to either totally or partially replace the components of conventional concrete. They include geopolymer, synthetic plastic, laterite, and natural fiber materials. These materials in most cases were waste agricultural, industrial, and mining materials before they are incorporated into mixtures to produce alternative concretes. The impregnation of these materials to form the alternative concretes is found to improve the mechanical strengths and durability of normal concrete in the same way steel-reinforced concrete has. The areas of concrete property improvements include tension, flexure, crack mitigation, toughness, impact resistance, acid, sulfate, and chloride attacks. Others like compression and alkaline attack may require the addition of admixtures for any significant improvements, while they generally may not improve workability and modulus of elasticity except further research efforts are invested.

The purpose of this manuscript is to report to construction stakeholders about the research, efforts, and practices of the past years on the alternatives to conventional concrete, and to show that alternative concrete from cheaper, lighter, and less hazardous components made largely from waste materials are available for use when the components are either fully or partially introduced to replace the conventional components of cement, sand, and gravel.

2. Alternative concrete composites

2.1 Cement and alternative concretes

Generally, cement or ordinary portland cement (OPC) is the most common type of cement around the world as a basic ingredient of concrete and mortar. This gray-colored (and sometimes white) powder material is a product of a fossil fuel energy-consuming manufacturing process with mainly limestone and some clay composites as raw materials. It is estimated that 2.8 billion tonnes produced annually in the world, and its production is responsible for about 8% of world carbon dioxide (CO_2) emissions [1, 2].

OPC is produced in a process called *calcination,* in which limestone (calcium carbonate) along with clay of silico-aluminous material is fed into a rotary kiln and heated at very high temperatures of about 1500°C. The calcination chemical equation is as given thus;

$$5CaCO_3 + 2S_iO_2 \rightarrow (3CaO.S_iO_2) + 5CO_2 \tag{1}$$

While the CO_2 gas is emitted into the atmosphere, the $3CaO.SiO_2$ (oxides of calcium and silicon, sometimes referred to as calcium silicates and oxides) compounds constitute the major constituent of cement. Minor constituents include aluminum oxide (Al_2O_3) and iron oxide (Fe_2O_3) that form part of the complex cement chemical composition. This $5CO_2$ exhaust in Eq. (1) combined with the fueling of the cement kiln is mostly responsible for its position as a major contributor to climate change.

When the cement is mixed with sand and gravel in the presence of water, it binds these components to produce the concrete composite. The hydration reaction that produces the binding gel is roughly described with the equations;

$$2C_3S + 6H \rightarrow C_3 - S_2 - H_3 + 3CH \tag{2}$$

$$2C_2S + 4H \rightarrow C_3 - S_2 - H_3 + CH \tag{3}$$

where C_3S and C_2S denote the calcium silicates and oxides, H denotes water, $C_3 - S_2 - H_3$ denote the calcium-silicate hydrate gel binder and CH denote calcium hydroxide.

The trouble with cement, at least in Nigeria, is not only about the greenhouse effect and the environmental pollution associated with its production but the high cost of procuring it due to the high prices that had kept increasing over the years. For example, **Table 1** shows the average nationwide prices in 5 years from market surveys, with the rates of increase that stood at 12.34% over the period.

The spiraling inflationary rate in the last 3 years is therefore a wake-up call for research and development in cheap and environmentally-friendly alternative materials.

2.1.1 Geopolymer concrete

Geopolymer concrete is a type of concrete in which the cement component is replaced with a material that is referred to as *Geopolymer cement* (GPC). In order to protect the environment, alternatives to OPC are required. Geopolymer cement is one of the most important alternatives though it may not be able to replace it completely. Credible research sources [4] have established that when cured at a temperature, GPC develops high strength and can be used for various applications.

GPC is an innovative and durable alternative binder to OPC. It is made from simply processed natural materials or industrial waste by-products. The natural materials include calcined clays, kaolinitic clays or metakaolin, lateritic clays, volcanic rocks, and mine tailings. The regular industrial waste by-products include ground granulated blast furnace slag (GGBS), which is a waste from Iron and steel production, and pulverized fly ash (PFA), which is a waste from coal and thermal plants. GGBS and PFA are rich in silica (oxides of silicon) and alumina (oxides of aluminum), which react with alkali activating solution by a process called *polymerization* to form molecular chains and

Year	2018	2019	2020	2021	2022	Ave. rate of increase (%)
Prices (NGN)	2,675	2,650	3,000	4,000	4,150	12.34
Rate of price increase (%)		−0.93	13.21	33.33	3.75	

Table 1.
Retail prices of cement for 50 kg bags in Nigerian Naira (NGN) [3].

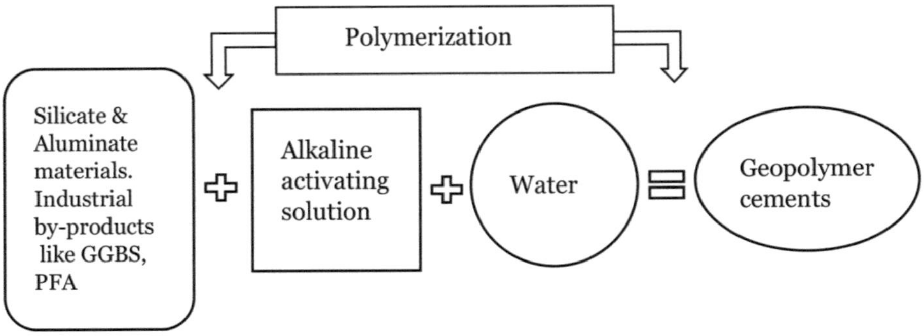

Figure 1.
The polymerization process.

networks of aluminosilicate gel that acts as the hard binding material for the concrete. This binding material is the GPC, and with this, the binding property serves as a complete or partial replacement for cement in concrete. The polymerization procedure is shown in **Figure 1**.

Extensive research elaboration has shown that the existing abundant raw materials resources, mainly industrial wastes, and the simple preparation technique that saves energy and environment accord GPC advantages over OPC. Also, the additional significant benefit of geopolymer cement represents their properties which include high early strength, low shrinkage, freeze–thaw resistance, sulfate resistance, and corrosion resistance. These properties make them the material advisable for application in various branches of industry besides the binder region [5].

The concrete so produced is referred to as geopolymer concrete. It is considered an eco-friendly alternative to conventional concrete because of the significant reduction of CO_2 footprint in producing the binding gel. The production process reduces emissions by 80–90%. Another attraction to GPC, making it gain importance and acceptance is that it ensures sustainability as most of the constituents are wastes that are recyclable. **Figure 2** is a diagram showing the benefits of geopolymer concrete.

Additionally, the improved engineering properties based on findings of research efforts confer some advantages on GPCs. These include the fact that they cure more rapidly than Portland-based cements and gain most of their strength within 24 hours to as much as 25 N/mm^2, and 70 N/mm^2 reported for 28 days [6]. However, they are set slowly enough so that they can be mixed at a batch plant and delivered in a concrete mixer. They also have the ability to form a strong chemical bond with all kinds of rock-based aggregates [7]. In some cases, GGBS replacements can be as high as 70% with 30% OPC for precast concrete manufacturing with reported increased resistance to chemical and chloride attacks, improved long-term strength gain, and lower heat of hydration enabling large volume pours due to reduced risk of cracking [8]. The use of newly developed accelerators as additives is making it possible to manufacture GGBS-based concrete without OPC with lower cost, increased strength, longer life, smoother finish, and white color it gives to the concrete. **Figure 3** shows a typical process of producing geopolymer concrete that has no OPC content.

2.1.2 Laterite concrete

Laterite is a type of soil found in hot and wet tropical and sub-tropical regions of the world, particularly in sub-Saharan Africa and Asia. It is formed from rocks that

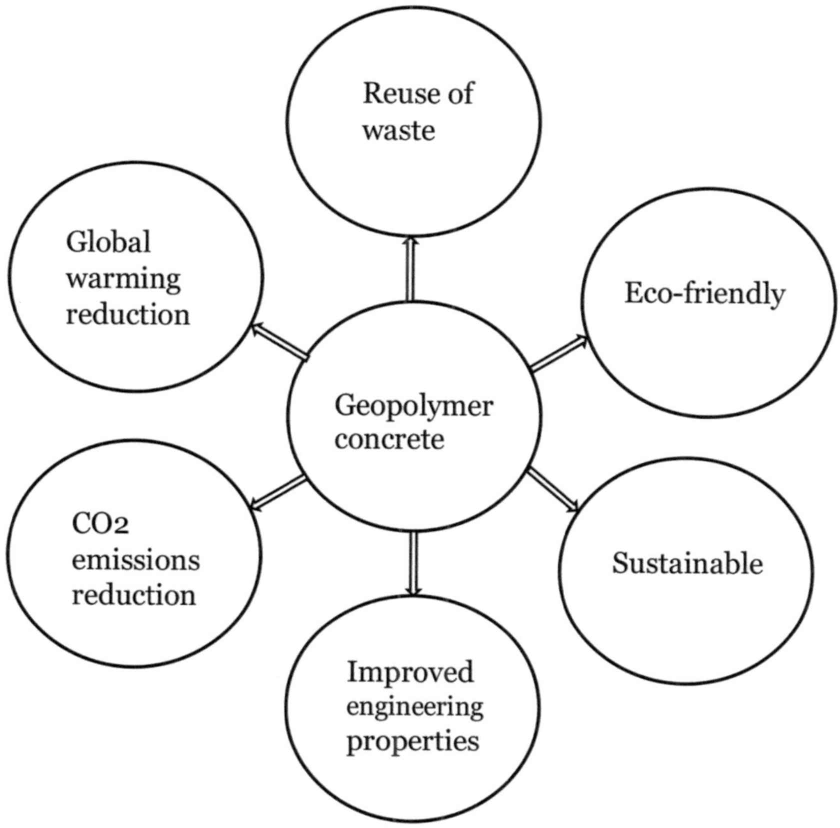

Figure 2.
Benefits of geopolymer concrete.

had weathered over many years under high temperatures and rainfall with wet and dry spells. Unlike geopolymer materials, it has none or very little silica content because the high rainfall leached away the silica and thereby leaving it rich in aluminates and iron oxides. The iron oxide constituent gives it various colors that range from red to brown and yellow. It becomes hard when exposed to the atmosphere, and its ease of abstraction makes its use in construction purposes desirable.

The laterite soil is a popular building material utilized in the regions of the world aforementioned because of its availability, relatively low – cost, and economical benefit compared to other natural earth materials. It is far cheaper than river sand used as fine aggregates in concrete and sandcrete building blocks and has no health hazard associated with silica dust since it leached away under tropical rainfall. In addition to its cost-effectiveness, laterite is also considered to possess better energy efficiency when compared to conventional modern building materials in the tropical regions of the world [9].

Laterite concrete or *laterized concrete* is a type of concrete, in which the fine aggregate of sand is partially or fully replaced with laterite soil. This alternative concrete composite has shown comparable engineering properties with the conventional concrete. It has been known to compare in strength with conventional concrete, though with less workability at fresh stage. This can be overcome with the use of admixtures. A review effort showed that laterite soil has proven to possess good

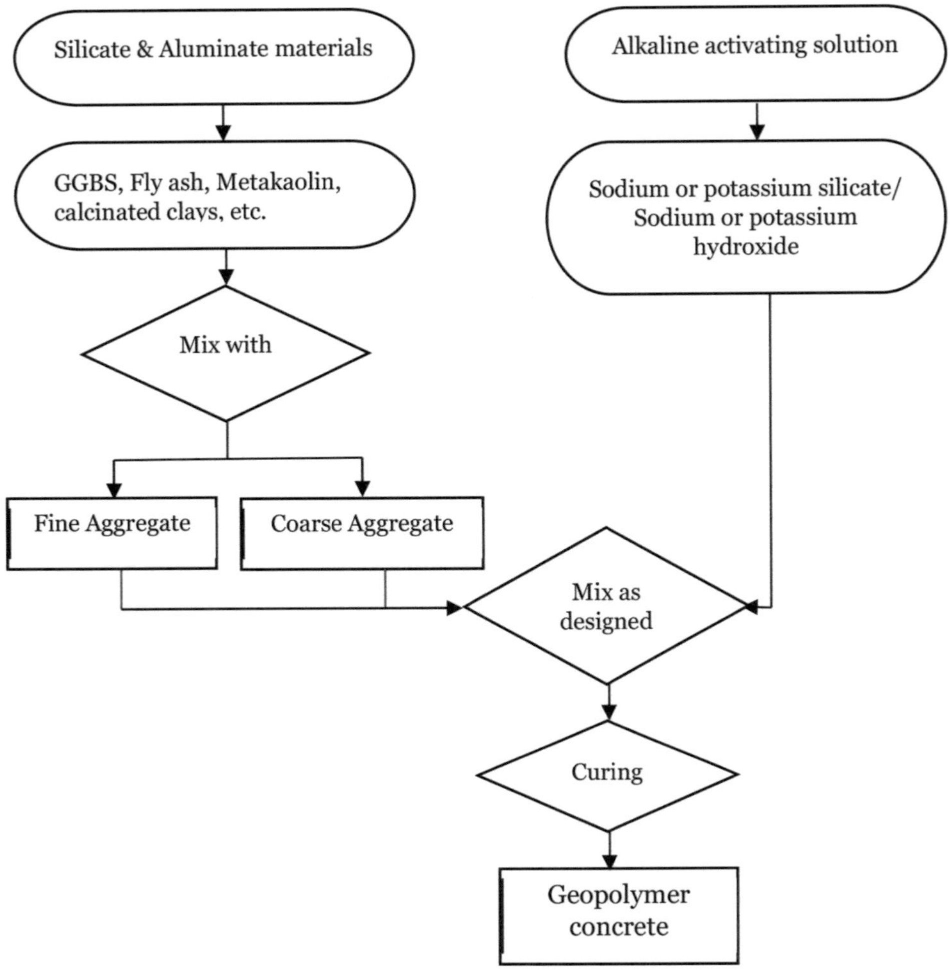

Figure 3.
Process of producing geopolymer concrete.

structural properties and can enhance some properties of concrete depending on the nature of the laterite and blended material [10].

An engineering consideration in the use of laterite in concrete among others is particle size distribution that cuts across the three main constituents of engineering soils, that is, clay (15–30%), silt (10–30%), and sand (40–75%). The gap-graded nature of river-washed sand, which is generally used in production of normal concrete, therefore accommodates it in a concrete mix. It is observed that better parking of aggregate materials within the concrete matrix due to a robust particle distribution resulted in enhanced shear resistance of laterite concrete beam.

According to a research report [11], the inclusion of laterite in concrete increased the shear resistance of concrete from 2.25 to 18.49 N/mm^2 for M20 concrete grade and from 5.17 to 13.14 N/mm^2 for M25 concrete grade at 15% replacement of river sand with laterite. **Table 2** shows some of the strength characterization results from concrete of M25 grade with improvements in compressive strength, flexural strength, density, and elastic modulus over the plain normal concrete (with 0% laterite). For

Sample tag	% Laterite	Compressive strength (N/mm^2)	Flexural strength (N/mm^2)	Density (kg/m^3)	Elastic modulus (GPa)	Shear strength (N/mm^2)
C0	0	22.27	9.883	2266.7	27.97	5.17
C5	5	27.40	9.925	2431.1	29.77	9.90
C10	10	21.24	9.870	2367.4	27.58	3.95
C15	15	26.32	12.693	2336.3	29.41	13.14
C20	20	24.46	10.800	2503.7	28.77	10.94
C25	25	19.79	9.199	2494.8	27.00	8.44

Table 2.
Some engineering properties of laterite concrete [9].

this grade of concrete, there is clear evidence of the increased toughness of the concrete beam, peaking at 15% replacement.

Since the compressive strength of concrete is considered a most important engineering property because of its basic function of resisting compressive load, laterized concrete has shown to provide adequate strength that equaled and exceeded 20 N/mm^2. Reports that show 40% sand replacements that produced 20 N/mm^2 [12] and a 100% laterite replacement with a 26.89 N/mm^2 strength [13] after 28 days of curing have been recorded.

Compressive strength results have generally ranged between 17 and 30 N/mm^2. This confers engineering viability on laterite concrete as a construction and building material.

Sandcrete blocks (i.e., often hollow blocks made from sand and cement) are very common building blocks. Because, just like concrete, they are not environmentally friendly and expensive to procure, *stabilized laterite blocks* are viable alternatives. Due to the relatively higher naturally occurring sand fraction in laterites, they bond well with cement in the presence of water to form strong and durable building blocks. Research results have shown that at 3–7% cement replacement of laterite, laterite-cement blocks improved in compressive strengths for up to 3.48 N/mm^2 after 28 days of curing, surpassing the minimum compressive strengths of 2.5 N/mm^2 specified by the national standards. Laterite concrete is also found to possess durability qualities, such as in water absorption capacity, surpassing sandcrete block at 4.88%, while the latter had 9.26% [9, 14, 15].

2.1.3 Pozzolana concrete

A pozzolana (or pozzolan) is a material with very high silica content or combination of silica and alumina that does not, in itself, have binding cementitious value or possesses a little of it but will, if finely pulverized into ash or dust form in the presence of water or moisture will chemically react with calcium hydroxide at room temperatures to form compounds that have binding and cementitious properties. Pozzolanas can both be naturally occurring or artificially made. Descriptions of various kinds of pozzolanas and their specifications are given in ASTM C618 and ASTM C1240 [16].

The reaction of silica (S) in pozzolana with calcium hydroxide (CH) produced from the reaction of OPC with water; it is a continuation of the OPC hydration reaction equation shown in Eqs. (2) and (3) and presented in Eq. (4);

$$Silica\ in\ pozzolana\ (S) + CH + H \rightarrow C - S - H \quad (4)$$

The resulting production of more calcium silicate hydrate ($C - S - H$) binder makes pozzolana an asset in concrete production. Because pozzolanas have to react with the calcium hydroxide produced from the hydration of OPC, partial replacement of OPC with a pozzolan then becomes perhaps the only relevant way of its use as a concrete/mortar material.

Natural pozzolanas are of volcanic origin. There are also *artificial* pozzolanas; this includes GGBS, pulverized fly ash (PFA), calcined clay, red brick dust and glass powder from industrial wastes, and agricultural wastes, such as rice husk ash (RHA), guinea corn husk ash (GCHA), and fonio husk ash (FHA).

These agricultural or post-harvest waste (sometimes referred to as *agro-waste*) materials that had constituted environmental nuisance over the years are found to exhibit pozzolanic properties when incinerated and converted to ash due to the very high silica and alumina content in the ashes. Other ashes include groundnut husk ash (GHA), millet husk ash (MHA), sugarcane bagasse ash (SCBA) and locust beam pod ash (LBPA), among others. An ash is considered a pozzolana according to ACI if the combined silica, alumina, and iron oxide components sum up to more than 70% of the oxide composition, that is, $SiO_2 + Al_2O_3 + Fe_2O_3 > 70\%$ [17].

An example of experimented ash analysis result that formed part of a concrete constituent is the GCHA shown in **Table 3** that has high content of silicate (78.192%) and the pozzolana criteria exceeding 70% at 80.374% from an atomic absorption spectrometer analysis [18].

Table 4 shows oxides composition for other ashes from agro-wastes. Ranges in values where they exist show different values from different authors. The slight differences are not unusual as varieties of crops are bound to differ in chemical combination ratios. The presented results show that all are pozzolanas and RHA along with GCHA stand out. The results also show that cheap and available agricultural wastes that constitute environmental nuisance and hazards over the years can readily be converted to concrete and mortar component materials for construction and building purposes.

In most of the results of research efforts on pozzolanas, the concretes and mortars made from the partial replacements of OPC with these ashes have resulted in appreciable strengths for the concretes depending on the percent replacement levels and the curing period. They set and harden more slowly than normal concrete, thereby making them more attractive and useful in mass concreting works. Usually, beyond 10% replacement, the strengths are reduced considerably. **Table 5** shows some results of the compressive strengths from various ashes.

In addition to competing strengths with cement and concrete composites made from it, the benefits of pozzolanas may be summarized as follows;

1. The economic gain obtained by partial replacement of OPC with cheaper pozzolanas, especially from ashes of agricultural wastes.

Oxide	**SiO_2**	**Al_2O_3**	**Fe_2O_3**	**K_2O**	**MgO**	**CaO**	**P_2O_5**	**SO_3**	**Others**
% Content	78.192	1.345	0.837	7.674	3.757	3.338	2.946	0.494	1.417
Total for Pozzolana		**80.374**>70%							

Table 3.
Oxides composition of GCHA.

Pozzolana	RHA (Range)	GCHA (Range)	GHA	FHA	SCBA
Oxide					
SiO_2	67.30–88.32	78.2–85.4	51.54	59.05	54.9
Al_2O_3	4.90	0–1.35	22.45	7.89	7.8
Fe_2O_3	0.95	0.64–0.84	2.40	3.12	10.0
SiO_2+ Al_2O_3+ Fe_2O_3	**73.15–94.17**	**78.84–87.59**	**76.39**	**70.06**	**72.70**
CaO	1.36	2.04–3.34	15.63	2.63	4.9
SO_3	2.70	0−0.49	0.94	0.30	1.2
Na_2O	—	0.25−0.98	—	0.90	0.2
K_2O	—	4.01−7.67	—	4.80	3.0
P_2O_5	—	0−2.95	0.60	0.80	—
MgO	1.81	0.01−3.76	1.20	1.73	2.5

Table 4.
Oxide composition of some pozzolanas from agro-wastes [19–25].

Pozzolana	RHA	GCHA	GHA	LBPA	FHA	Neem seed Ash concrete
% Replacement of OPC						
0	25.5−41.0	22.9–25.5	21.8	22.2	25.1–26.1	40.1
5	34.6	23.5–26.3	—	20.9	27.0	—
10	26.1−28.1	23.51	18.4	13.7	20.0–25.6	41.9

Table 5.
Compressive strength of concretes from some agro-waste pozzolanas in N/mm^2 [21, 24, 26–30].

2. Massive reduction of negative environmental implications associated with the carbon dioxide gases emitted during OPC production and concrete manufacture.

3. Improved durability of the concrete and mortar made from pozzolanas without reducing the compressive strength or other characterized performance indices in a significant way. The durability indices include water absorption, permeability, and resistance to aggressive environments that have sulfates, acids, and chlorides.

4. The recycling process of converting wastes into durable building materials and thereby enhancing environmental sustainability.

2.1.4 Polymer concrete

Polymer concrete (PC) is a special type of concrete in which polymer is used as an alternative to OPC as a binder. In some cases, the polymer may be used along with OPC to make the composite, which is referred to as polymer cement concrete (PCC) or polymer modified concrete (PMC). Polymers are used for making shopping bags, water bottles, vulcanized rubbers, food packaging, and auto parts. Their tendency to end up as after-use waste materials makes them attractive as cheap and environmentally sustainable materials for building purposes.

To produce the PC, the polymer or monomer is mixed with fine and coarse aggregates under heating, then placed in position before it is polymerized. Polymerization can be carried out by a number of means. This includes the methods of elevated thermal catalytic reaction, catalyst promoter reaction, or radiation.

The American Concrete Institute (ACI) [31] postulates that for a composite to qualify as PC, it must possess a unique combination of properties. The properties include fast curing at temperatures between −18 to +40°C; high strengths in tension, flexure, and compression; good adhesion characteristics on surfaces; durability over a long period of time through low permeability to water and aggressive solutions and thawing and freezing cycles and chemical resistance, in addition to being lightweight. The durability properties of corrosion resistance and impermeability of PC an advantage in its use in building build swimming pools, making sewerage and drainage channel systems, and other structures containing liquids or corrosive chemicals.

The compressive strength obtained with PC can be as high as 140 N/mm^2 within a curing period that is short. However, these characteristic properties depend on the type and quantity of polymers used in the concrete. Types of polymers include polyester resin, which is used in construction, such as in fiberglass, general waterproofing, and repairs; vinylidene chloride, which is used to make plastics for food wraps and in packaging; epoxy; styrene, which is used as a strong adhesive in automobiles and aircrafts, among others; and acrylics, which are used in makeups, particularly with nails.

The advantages of PC as an alternative concrete against the normal conventional OPC concrete include the following:

1. It possesses better compressive, flexural and tensile strengths, and abrasion resistance.

2. It has a very good adhesion on most surfaces and can be applied in very thin cross-sections.

3. It can attain up to 70% strength after 24 hours of curing at room temperature, therefore requiring fast treatment, whereas OPC concrete may attain 20% strength after 28 days.

4. It offers long-term durability as earlier discussed, reduces shrinkage, and has lightweight.

5. It reduces the infiltration of carbon dioxide and protects the PC from carbonation.

6. It is useful in repairing existing structures due to the fast-setting binder and its resistance to weathering effects.

It also has the following disadvantages:

1. It costs more to produce, however, this is ameliorated or minimized if the polymer is an after-use waste material.

2. It requires very high skills and precision work during production, especially in proportioning and mixing.

3. It is very important to make use of masks and hand gloves for skin safety because of the type of dangerous chemicals used in PC production.

2.1.5 Expanded polystyrene (EPS) concrete

The expanded polystyrene (EPS) material (also called Styrofoam) is a rigid, tough, hydroscopic, and light-weight foam made from a pre-expanded polystyrene bead. It is used as food containers, insulations in buildings, and for packaging of fragile items inside box containers. According to some research and industry sources [32, 33] the world produces over 14 tons of polystyrene every year, much of which after use is often discarded as no-cost domestic and industrial waste materials. Since they are non-degradable, they constitute environmental hazards and nuisance in landfills, if not converted for good.

EPS concrete (EPSC) or EPScrete is made from cement and balls of EPS with little or without aggregates. it is not usually as strong as normal concrete. Its possible preference over normal concrete is based on its lightweight, excellent heat preservation, and sound insulation. Others, according to some findings [34] are its hydrophobic properties, lower cost, environmental friendliness and increased thermal insulation properties. It is increasingly being used in various applications in constructing environmentally "green" homes. The global green building movement, an already huge factor in the real estate development sector for residential housing is at the forefront of this.

2.1.6 Fiber reinforced concrete (FRC)

Fiber-reinforced concrete (FRC) is an alternative and cost-effective composite material to conventional steel-reinforced concrete (CSRC). It usually consists of fibrous materials that improve the structural integrity of the concrete or mortar. FRC may also be described as concrete that contains slender, and most times, short discrete fibers that are uniformly distributed and randomly oriented in the concrete mixture for the purpose of improving the structural performance. The fibers can be obtained in natural or synthetic forms with each providing varying properties to the concrete.

Important advantages of FRC apart from cost-effectiveness include time-saving as a result of adding fiber and mixing offsite. When this ready-to-use concrete is delivered, it allows for faster placement and eliminates handling time in contrast with steel-reinforced concrete works, where a lot of effort and time are expended in bar scheduling, bar bending, and bar installation. The solution provided by FRC also has the potential to reduce carbon dioxide emissions when compared to the provisions of reinforcing steel bars.

The general attraction of fibers in concrete is to control cracking due to plastic and drying shrinkages. Others are to reduce the permeability of concrete leading to reduction of the bleeding of water and increased shear and strain capacities.

The fibers in FRC can be classified into two categories; *Microfibers* and *Macrofibers.* While the earlier are usually shorter and thinner, the latter are longer and may be thicker. In some constructions, both may be combined to form composite that includes mixtures of cement, mortar, or concrete and discontinuous, discrete, uniformly dispersed suitable fibers for improved properties. In other cases where the load bearing is very high, they may be combined with the regular steel bar reinforcements.

Micro fiber reinforced concrete (MiFRC) has many individual fibers that are dispersed throughout the concrete as evenly as possible during the mixing process. The matrix-like structure so created provides long-term durability to concrete while reducing the frequency of plastic shrinkage cracks in fresh concrete after placement. Another benefit is the tougher surface it possesses, which makes it have more resistance to impact stress.

Macro fiber reinforced concrete (MaFRC) has individual fibers arranged and spread in the concrete matrix. They improve the performance characteristics of the concrete by increasing toughness and ductility of hardened concrete. They also improve post-crack performance by allowing concrete to retain the load-carrying capability after cracking has occurred. Other general benefits of macro fibers in concrete include increased tensile and flexural strengths, improved impact and abrasion resistance.

Over the years MiFRC and MaFRC have found engineering applications in the construction of floors, car parks, airports, bridges, domestic, and agricultural infrastructure projects. They are also applied in places and instances when conventional reinforcement placement is very difficult or not possible like in very thin structures.

2.1.7 Types of fiber-reinforced concrete (FRC)

The major factors that generally affect the performance characteristics of FRC are the water/cement ratio, the quantity or percentage content, diameter, and length of fibers or the aspect ratio. **Figure 4** gives the different types of FRCs used in construction with brief discussions of them.

2.2 Steel fiber reinforced concrete (SFRC)

The steel fiber is a product of steel metal of a specified geometry with an aspect ratio (ratio between length and diameter) between 20 and 100. Steel is one of the most important materials in the construction industry. It provides tensile resistance to concrete since concrete is known to be weak in tension and brittle. Therefore, the presence of its fiber in concrete can cause changes in the quality of the concrete's physical property by greatly increasing resistance to tension, cracking, impact, fatigue, bending, and durability. It also improves other mechanical strengths like toughness and stress resistance in concrete.

Compressive strength of SFRC is affected mainly by the quantity of fibers, fiber sizes, and bonding with the matrix. Micro fibers mitigate the development of microcracks in the concrete, resulting in improved compressive strengths. On the other hand, macro fibers help to control crack openings, which increases the concrete energy absorption capacity. However, increased fiber addition may cause some distress in the matrix, which can result in creation of more voids. The designer, therefore, will need to create a balance that aims at an optimum fiber quantity and fiber sizes for comparable compressive strength values. Previous research results [35–37] show that SFRC significantly improved in flexural and axial strengths with overall toughness compared to the normal reinforced concrete. Other properties like compressive and shear strength increased at up to 1.0–1.5% optimum fiber content by weight after which there was reduction in strength, impact resistance improved up to

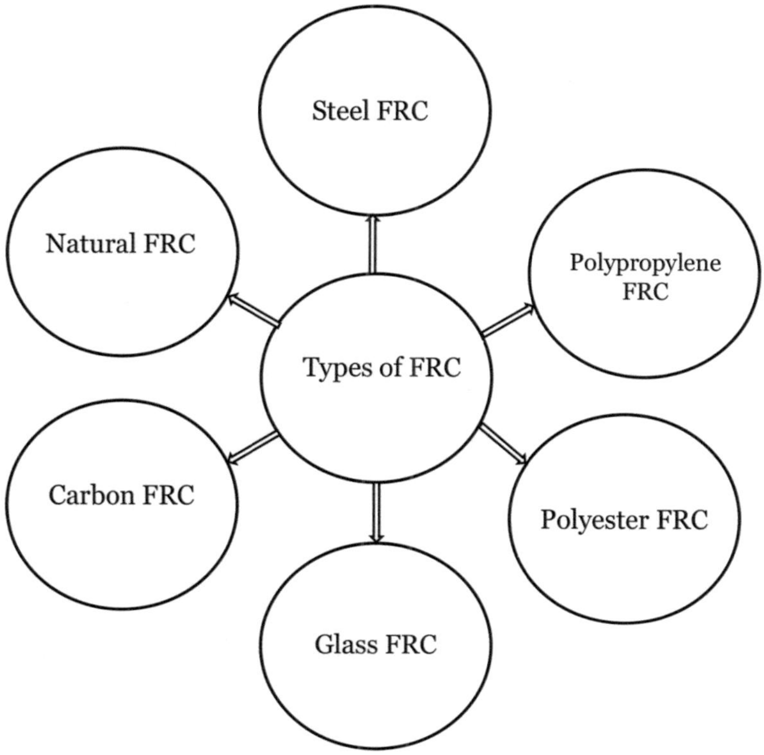

Figure 4.
Types of fiber reinforced concrete (FRC).

10 times over plain concrete and there was higher resistance to corrosion. However, modulus of elasticity decreased with an increase in fiber content.

SFRC has found structural applications in the design and construction of floors, buildings, precast concrete, concrete bridges, tunneling, and heavy-duty rigid pavements. The types of steel fibers used in SFRC are specified by ASTM A820/A820M [38] which for example, specifies that the average tensile strength of fiber should not be less than 345 MPa. It also has specifications on fiber dimensions and workmanship, among others.

2.3 Polypropylene fiber reinforced concrete (PpFRC)

Polypropylene Fiber (PpF) is a synthetic fiber made from propylene with similar properties to polyethylene, though harder and more heat resistant. It displays good heat-insulating properties, is highly resistant to acids, alkalis, and organic solvents, and is used in a variety of applications.

Research results [39] have shown that adding PpF to concrete improved its compressive and tensile strengths without regard to the water-cement (w/c) ratio. For example, adding 0.5% of PpF by volume to concrete with w/c ratios of 0.46 and 0.36 increased compressive strengths by 12 and 6%, respectively, while for tensile strengths it increased by 17 and 8%, respectively. The impact strengths increased 5 and 2.1 times at first cracks for the two w/c ratios, respectively, which translates to higher energy absorption capacity in the concrete.

Also, in a hybrid FRC involving the combination of steel and PpF at constant 0.75% by volume [40], an attractive solution for enhancing the post-cracking behavior of concrete was found in compression and flexure though without a significant effect on their strength values. The even compressive and bending strength values might have been due to the relatively low fiber percent in the mix as optimum fractions from many works stand at 1.0–1.5%. This is further attested to by Blazy and Blazy [41], where it was generally mentioned that PpF improves concrete properties, but until a certain dosage (i.e., optimum value) after which if exceeded will have reduced performance.

In summary, the effect of PpF on the properties of concrete can be described as *very improved* in toughness, crack mitigation and limitations, impact and spalling resistance, anti-freeze–thaw cycles and durability; *improved* in flexural and tensile strengths, abrasion resistance, eco-friendliness, and economics; *requiring further assessment* in water absorption, porosity, and permeability; *slightly neutral* in compressive strength and modulus of elasticity and *totally neutral* in workability.

2.4 Polyester fber reinforced concrete (PeFRC)

Polyester fiber is a synthetic polymer that is manufactured when there is a chemical reaction between an acid and alcohol. It is usually presented as non-biodegradable industrial waste. The raw materials are usually petroleum, coal, water, and air. It has appreciable elasticity, toughness, and sound absorption properties. It is traditionally used in the textile industry and sometimes blended with cotton to produce clothing.

Research results [42] on PeFRC indicate follows: an increase in impact resistance (166.7%); an increase in tensile (65.5%), flexural (66.7%), and compressive (21.5%) strengths; and decreases in workability and modulus of elasticity (7.14%). The above results came from optimum fiber content of 0.075 and 10% cement replacement with fly ash. However, in another result without fly ash component [43], the, tensile strength, flexural strength, and compressive strength of the PeFRC increased by 24.3, 15.13, and 36.2%, respectively compared to normal OPC concrete after 28 days of curing with an optimum fiber content value steadying at 3.5%. This shows that admixtures have positive effects on the strength and the quantity of fibers used in concrete by increasing strength and reducing fiber content.

Other results in Ref. [44] indicate that in addition to flexural and compressive strengths improvements, there was resistance to abrasion, alkali resistance and reduced drying shrinkage when compared to normal OPC concrete.

The fiber is used to reinforce concrete which has found applications in the construction of industrial and warehouse floors, pavements, and precast concrete. Like with the other FRCs they provide adequate resistance to the formation of plastic shrinkage cracks and enhance toughness.

2.5 Glass fiber reinforced concrete (GFRC)

Glass fiber compares well with polymers and carbon fiber in mechanical strength. It is much cheaper and less brittle when used in composites. These characteristics make it useful as a reinforcing component in many polymer composite products. The resulting products called glass-reinforced plastic (GRP) or *fiberglass* are very strong and relatively lightweight. However, when, rather than reinforcing polymer with glass, OPC and sand are used to replace polymer, then, the resulting material will be GFRC.

Compressive review of glass fiber in concrete [45] shows that it impacted negatively on workability of GFRC with an increase in fiber content. While it enhanced the tensile and flexural strengths, it did not significantly improve the compressive strength within the optimum fiber contents of 1–2%. It resisted cracking but did not significantly improve surface hardness. It has significant improvement in durability properties by reducing permeability of GFRC to chloride-ions because of its ability to mitigate against cracks; it improved performance in marine environments under acid and sulfate exposures. It is suggested that further research that incorporate pozzolanas be investigated to improve compressive strength of GFRC in view of the importance of this property in load bearing function of concretes.

GFRC has been used for many years to produce different and varied products.

The products include thin architectural cladding panels, ornamental concrete, such as domes, statues, fountains, decorative panels, concrete countertops, and artificial rock work.

The challenge with using glass fibers in GFRC is that since glass is primarily a silica material it breaks down in the alkaline concrete environment due to alkali-silica reactivity (ASR) when there is reactive silica in the concrete aggregate. However, innovations in alkali-resistant (AR) glass fibers including impregnation of pozzolanas have helped in overcoming the alkaline attack leading to a rapid increase in GFRC use.

2.6 Carbon fiber reinforced concrete (CFRC)

Carbon fibers are mostly made up of carbon atoms that in physical appearance are thinner than even human hair, measuring about 5–10 micrometers in diameter. They can be combined to form a yarn or processed to make a grid structure and then coated. The benefits of carbon fibers include high stiffness, high tensile strength, lightweight, high chemical resistance, high-temperature tolerance, and low thermal expansion. When compared to steel, carbon fiber is four times lighter and yet possesses six times more load-bearing capacity. When the fiber is integrated into a plastic resin under heat it becomes a composite called carbon fiber reinforced polymer or commonly referred to as carbon fiber, possessing very rigid property and high strength/weight ratio, though brittle.

Carbon fiber reinforced concrete is an innovative combination of carbon fiber fabrics or bars with fine-grained concrete composite material. This type of concrete can be formed into varied shapes including slimmer and delicate structures and yet with a high load-bearing capacity. CFRC can be used for repairs of existing structures and for sustainable less- material-intensive and lightweight construction.

According to a research result [46], a concrete mix with 0, 0.5, and 0.75% carbon fiber contents produced compressive strengths of 40, 45 and 48.7 N/mm^2, respectively, and flexural strengths (measured as modulus of rupture) of 4.8, 5.9 and 6.1 N/mm^2, respectively. The results were determined after 28 days of curing and admixtures were added to the concrete mix. These results compare significantly M40 concrete in compression and exceed considerably the flexural values of the same grade of concrete.

In another research output [47], it was found that by including carbon fibers in concretes of different compressive strengths, the fibers provided for the concrete of low strength (M20 grade) an efficiency factor of 258.6, the medium strength (M30 grade) had an efficiency factor of 62.1 and for high strength (M40 grade) concrete it provided 2.7 as efficiency factor. What this implies is that carbon fiber not only improves strength but also can provide the most effective reinforcement for the concretes of low strength.

Yet in related work [48], fiber content was varied at 0, 0.75, 1, and 1.25% in an M25 concrete and though workability decreased, compressive strength increased by 46.8, 59.9, and 32.4% respectively with optimum content value at 1%. It was also observed that split-tensile strength increased at 28.1, 56.3, and 9.4% over values of normal concrete again at 1% optimum fiber content. The values for flexural strength increased by 88.5, 107.69, and 78.46% respectively at 1% optimum content. The same result showed that CFRC resisted acid and sulfate attacks far better than normal concrete, especially at 1% optimum fiber content. The use of GFRC as an alternative to conventional concrete is well established.

2.7 Natural fiber reinforced concrete (NFRC)

Synthetic fibers manufacture is relatively expensive and their production process demands considerable energy. An alternative is the use of low-modulus natural fibers in the production of building composites. Natural fibers are usually cheap, available in many parts of the developing world as waste materials constituting environmental nuisance which can be deployed for sustainable use as building material components.

Natural fibers are made up of combination of cells in which the thickness is almost negligible when compared with the length, that is, with a high aspect ratio. They can broadly be classified as being from vegetable or animal origins. Fibers of vegetable origin include coconut husk or coir, sisal, sugarcane bagasse, bamboo, jute, palm fruit husk, wood fibers, and straw, and those of animal origin include silk and wool. Natural fibers of vegetable origin have a huge history of use as reinforcing and stabilizing materials for soil-based composites like mud, clay, and laterites to improve the properties of bricks, roofing and plaster materials in buildings [49].

Natural fibers are potential choices for reinforcing concrete because of their availability and cheapness, particularly in developing countries, and improvement in mechanical properties. A research result [50] on the use of sisal fibers found that improvements in fresh concrete and hardened concrete with significant increases were recorded for compressive strength at optimum fiber content of 1.5%, split tensile strength increased even at 0.5% fiber content, and flexural strength improved at 1% fiber content over normal concrete.

A comprehensive review of coir, sisal, jute, and bamboo fibers in concrete [51] shows that the usage of the fibers improved the concrete's compressive, tensile, and flexural strengths, durability, and the load-carrying capacity. Another finding was that coir and sisal not only delayed and controlled cracking due to tension but presented the best improvements in the concrete properties with optimum fiber contents averaging 1.5%. However, admixtures like micro silica assisted in strength and durability enhancements.

In another study [52], the conclusions were similar to the ones mentioned above with enhanced fracture toughness, impact strength and crack resistance except that there was no significant increase in compressive strength values and workability values were higher. These results are good, economical, and beneficial in rural communities where the fibers are freely and easily available.

The use of coir in cement mortar and mathematical optimization of the mortar mixture was investigated [53] and the results confirmed the superiority of the fiber mortar over plain mortar with the optimized compressive strength exceeding the laboratory value by 42.8%.

A major disadvantage with natural fibers in concrete is that as biodegradable materials they are susceptible to decay over time resulting in to loss of strength and subsequent brittle failure. However, this can be avoided by pre-treatment of the fibers with resins and gums before mixing and inclusion of admixtures like pozzolanas (rice husk ash, silica fume, etc.).

3. Conclusion

It is established that research efforts and practices over the years have found credible alternatives to conventional concrete and reinforced concrete, as highlighted in this chapter. The general benefits include the following:

1. Alternative concretes produced from geopolymers, laterites, pozzolanas, polymers, expanded polystyrene, and fibers constituents possess adequate mechanical strengths and durability qualities respectively to compete with conventional plain concrete and even steel bar reinforced concrete.

2. These alternative concrete materials are mostly derived from industrial, agricultural, mining, and domestic wastes thereby making them very cheap, readily available, and economical. They also make for environmental sustainability when used for building and construction purposes.

3. The use of these materials over cement and steel will reduce the incidence of carbon dioxide emissions associated with the manufacture of cement and steel and thereby curtail negative climate change.

4. Since concrete is brittle and very deficient in tensile strength, these alternative concretes have significantly provided improved tensile and flexural strengths for concrete, and compressive strength to some extent. Other property enhancements include impact resistance, toughness, cracks mitigation; acid, chloride, sulfate resistance, and other durability properties.

5. In addition to competing use in conventional constructions, these alternative concrete have especially found applications in the construction of thin delicate structures often of architectural importance, in the repairs of existing structures, and in mass concreting of heavy-duty floors.

6. Nevertheless, further investigations are suggested in improving their workability, compressive strength, alkaline attack in glass fiber reinforced concrete, permeability, and water absorption, though these properties are presently being treated with incorporation of admixtures.

Conflict of interest

The author declares no conflict of interest. All the figures were made by the author and so did not warrant copyright permission.

Author details

Emmanuel Ndububa
Department of Civil Engineering, University of Abuja, Abuja, Nigeria

*Address all correspondence to: emmanuel.ndububa@uniabuja.edu.ng

References

[1] Gagg CR. Cement and concrete as an engineering material: An historic appraisal and case study analysis. Engineering Failure Analysis. 2014;**40**:114-140

[2] Timperley J. If the cement industry were a country, it would be the third largest emitter in the world. Cement & Its Environmental Impact. 2021;7:1

[3] Ayemba D. The price of cement in Nigeria. Construction review online. 22 May 2022

[4] Singh NB, Mukesh K, Sarita R. Geopolymer cement and concrete: Properties. Materials Today: Proceedings. 2020;**29**(3):743-748

[5] Zivica V, Palou MT, Kvizma M. Geopolymer cements and their properties: A review. Building Research Journal. 2014;**61**(2):85-100

[6] Madhavi TC, Rameshwaran PM. Geopolymer concrete - The eco friendly alternate to concrete, NBM&CW infra construction & equipment magazine. May 2020

[7] Davidovits J. Geopolymer cement, a review. Geopolymer Science and Technics. 2013;**21**:1-11

[8] Stuart J. The PH+ guide to greener concrete- reducing the climate impact of cement and concrete in buildings, passive house + sustainable building, no. 33. 2020

[9] Ndububa EE. Stabilized lateritic bricks as alternative to mud housing in Bauchi, North East Nigeria. IOSR Journal of Mechanical and Civil Engineering (IOSR-JMCE). 2017;**14**(5):67-73

[10] Shuaibu RA, Mutuku RN, Nyomboi T. A review of the properties of laterite concrete. International Journal of Civil and Structural Engineering. 2014;**5**(2):130-143

[11] Ndububa EE, Ocholuje SO. A computational assessment of the deflection and shear improvement of concrete beam impregnated with laterite. Asian Journal of Civil Engineering. 2022;**23**:173-186

[12] Udoeyo FF, Iro UH, Odim OO. Strength performance of laterized concrete. Construction and Building Materials. 2006;**20**(10):1057-1062

[13] Ettu IO, Ibearugbulem OM, Ezeh JC, Anya UC. The suitability of using laterite as sole fine aggregate in structural concrete. International Journal of Scientific & Engineering Research. 2013; **IV**(5):502-507

[14] NIS. Nigerian Industrial standard for sandcrete blocks. In: Nigerian Industrial Standard. Lagos: The Standard Organization of Nigeria (SON); 2004

[15] Aderibigbe YW, Ataguba OC, Abuh NM, Ichado J. Assessment of properties of sandcrete blocks produced by commercal block industries in Idah, Kogi State, Nigeria. International Journal of Scientific Engineering and Science. 2017;**I**(3):8-11

[16] A. S. A. T. Measurements, ASTM C1240-20 Standard Specification for Silica Fume used in Cementitious Mixtures. West Conshohocken, PA: ASTM International; 2020

[17] A. S. A. T. Measurements, ASTM C618–22 Standard Specification for Coal Fly Ash and Raw or Calcined Natural Pozzolan for Use in Concrete. West Conshohocken, PA: ASTM International; 2022

[18] Ndububa EE, Nurudeen Y. Effect of guinea corn husk ash as partial

replacement for cement in concrete. IOSR Journal of Mechanical and Civil Engineering (IOSR-JMCE). 2015;**12**(2): 40-45

[19] Aburime PI, Ndububa EE, Kpue DI. The impact of Guinea corn husk ash as an admixture for crack control in concrete. European Journal of Technology Research. 2020;**5**(10):21-31

[20] Ndububa EE, Aburime PI. Flexural strength of concrete made from guinea corn husk ash as partial replacement for cement. Nigerian Journal of Engineering. 2021;**28**(1):25-31

[21] Manasseh J. A review of partial replacement of cement with some agro wastes. Nigerian Journal of Technology. 2010;**29**(2):12-20

[22] Habeeb GA, Hilmi BM. Study on properties of rice husk ash and its use as cement replacement material. Materials Research. 2010;**13**(2):185-190

[23] Ketkukah T, Ndububa EE. Ground nut husk ash (gha) as a partial replacement of cement in Mortar. Nigerian Journal of Technology. 2006; **25**(2):84-90

[24] Ndububa EE, Okonkwo JS, Ndububa OI. The potential use of fonio husk ash as a Pozzolana in concrete. Nigerian Journal of Technology. 2016; **35**(1):31-36

[25] Chindaprasirt P, Rattanasak U. Eco-production of silica from sugarcane bagasse ash for use as a photochromic figment filler. Scientific Reports. 2020; **10**(1):1-8

[26] Ibrahim MH, Ramadhansyah PJ, Rosli MH, Ibrahim WH. Mechanical performance of porous concrete pavement containing nano black rice husk ash. In: IOP Conference Series: Materials Science and Engineering. 2018

[27] Ndububa EE, Uloko JO. Locust bean pod ash (LBPA) as a ppzzolanic material in concrete. In: Proceedings of The International Academic Conference for Sub-Sahara Africa Transformation and Development, Ilorin. 2015

[28] Olutoge FA, Adesina P. Effects of rice husk ash prepared from charcoal-powered incinerator on the strength and durability properties of concrete. Construction and Building Materials. 2019;**196**:386-394

[29] Marfo SA, Owusu WA. Evaluation of Guinea corn husk ash as oil-well cement slurry extender. International Journal of Petroleum and Petrochemical Engineering (IJPPE). 2018;**4**(4):1-7

[30] Aishwaryalakshmi V, Keerthi P, Abinaya B. Experimental study on partial replacement of cement by neem seed husk ash in concrete. International Journal of Civil Engineering and Technology (IJCIET). 2018;**9**(11): 2739-2744

[31] A. C. I. ACI. CT-13: ACI Concrete Terminology - An ACI Standard. Farmington Hills, MI: ACI Foundation; 2017

[32] Herki BA, Khatib JM. Valorisation of waste expanded polystyrene in concrete using a novel recycling technique. European Journal of Environmental and Civil Engineering. 2017;**21**(11): 1384-1402

[33] Khatib JM, Herki BA, Elkordi A. Use of Recycled Plastics in Eco-efficient Concrete: Characteristics of Concrete Containing EPS. Amsterdam: Elsevier Ltd. and Woodhead Publishing Series in Civil and Structural Engineering; 2019. pp. 137-165

[34] Bhutta MA, Yoshihiko O, Tsurata K. Strength properties of polymer mortar panels using methyl methacrylate solution of waste expanded polystyrene as binder. Construction and Building Materials. 2011;**25**(2):779-784

[35] Hamid BP, Behzad N, Majid F. Steel fiber reinforced concrete: A review. In: Proceedings of the International Conference on Structural Engineering Construction and Management (ICSECM), Kandy-Sri Lanka. 2011

[36] Neves DR, Almeida JC. Compressive behaviour of steel fibre reinforced concrete. Structural Concrete. 2005;**6**(1): 1-8

[37] Nafisa T, Jonayed KS, Saiful I, MoinulIslam I. A study on the strength characteristics of steel fiber reinforced concrete. In: 2nd International Conference on Advances in Civil Engineering, Chittagong. 2014

[38] ASTM. A820/A820M Standard Specification for Steel Fibers for Fiber-Reinforced Concrete. West Conshohecken, MI: ASTM International; 2011

[39] Alavi NA, Hedayatian M, Nili M, Afrough SV. An experimental and numerical study on how steel and polypropylene fibers affect the impact resistance in fiber-reinforced concrete. International Journal of Impact Engineering. 2012;**46**:62-73

[40] Antonio C, Serena G, Enzo M, Nicola N, Marco P. Experimental characterization of the post-cracking response in hybrid steel/polypropylene fiber-reinforced concrete. Construction and Building Materials. 2016;**145**: 1035-1043

[41] Blazy J, Blazy R. Polypropylene fiber reinforced concrete and its application in creating architectural forms of sublic spaces. Case Studies in Construction Materials. 2021;**14**:1-19

[42] Tharun AP, Nishma MV, Aswathy LS, Sruthy S, Aparna AV. Strength characteristic study of polyester fiber reinforced concrete. International Journal of Engineering Research & Technology. 2018;**XI**(6):1-6

[43] Venu M, Neelakanteswara PR. Polyester fibers in the concrete an experimental investigation. Advanced Materials Research. 2011;**261-263**: 125-129

[44] Gupta S, Rao KV, Sengupta J. Evaluation of polyester fiber reinforced concrete for use in cement concrete pavement works. Road Materials and Pavement Design. 2008;**IX**(3):441-461

[45] Ahmad J, Gonzalez-Lezcano RA, Majdi A, Ben Kahla N, Deifalla AF, El-Shorbagy MA. Glass fibers reinforced concrete: Overview on mechanical. Durability and Microstructure Analysis, Materials. 2022;**XV**(5111):1-23

[46] Aljalawi NM, Al-Jelawy HM. Possibility of using concrete reinforced by carbon fibre in construction. International Journal of Engineering & Technology. 2018;**XII**(4.20):449-452

[47] Branco LA, Aguiar LC, Chahud EC, Lahr FA, Panzera TH, Christoforo A. Reinforcement of concretes with carbon fiber composite. International Journal of Composite Materials. 2014; **IV**(2):63-68

[48] Navya HA, Nayana NP. Experimental studies on behaviour of carbon fiber reinforced concrete. International Journal of Civil Engineering and Technology. 2018;**9**(7): 1461-1469

[49] Ndububa EE, Yusuf MI. Strength improvement of mud houses through stabilization of the lateritic materials. The International Journal Of Engineering And Science. 2016;**5**(9):56-60

[50] Balasubramanian JC, Senthil SS. Experimental investigation of natural fiber reinforced concrete in construction industry. International Research Journal of Engineering and Technology. 2015; **II**(1):179-182

[51] Shadheer AM, Ravichandran P, Krishnaraja AR. Natural fibers in concrete – A review. In: IOP Conference Series: Materials Science and Engineering. 2021

[52] Kavitha S, Felix Kala FT. A review on natural fibres in the concrete. International Journal of Advanced. Engineering and Technology. 2017;**1**(1): 32-35

[53] Ndububa EE, Osadebe NN. Quartic modelling and optimization for the compressive strength of fibre cement using sceffes simplex lattice. NSE Technical Transactions. 2016;**50**(1): 70-82

Chapter 4

Creep in Concrete

Winfred Nthuka Mutungi

Abstract

Creep in concrete is a long-term deformation under sustained loading. It's influenced by many factors including constituent materials, environmental conditions among others. Whenever there is an alteration in the convectional concrete preparation process, the creep characteristics need to be realistically assessed. Creep prediction models have been developed for determining creep in convectional plain concrete. It has been shown that creep changes when constituents of concrete production are changed hence realistic assessment needs to be done. Creep of some modified concrete have been presented in this chapter.

Keywords: concrete, creep, creep prediction model, modified concrete, mechanism of creep

1. Introduction

1.1 Concrete definition

Concrete is one of the most used building materials. It is a composite material made from readily available constituents that is aggregates, sand, cement, and water. Concrete is a versatile material and can be mixed to meet a variety of special needs and formed to virtually any shape. There has been a growing demand of concrete worldwide with 27 billion tonnes of concrete produced annually and the amount expected to be four times in 2050 than it was in 1990 [1]. This has put pressure on the existing natural resources for its production and threatening their depletion. It has also raised environmental concerns on sustainability of concrete production by convectional materials.

With the above concerns, research has been undertaken on alternative ingredients for production of concrete. Many industrial wastes have been studied and found that some possess similar properties to the constituents of concrete. These industrial wastes have been recycled as substitutes of either cement or aggregates. They include fly ash slag, ground granulated blast furnace slag (GGBS), waste glass and even ground vehicle tires among others. Thus, concrete production can significantly reduce environmental impacts of industrial waste whilst improving the properties of concrete.

1.2 Creep definition

Creep is defined as increase in strain under a sustained constant stress after considering other time-dependent deformations not associated with stress e.g., shrinkage and swelling. When hardened concrete is loaded, it deforms; partly because of elastic

 IntechOpen

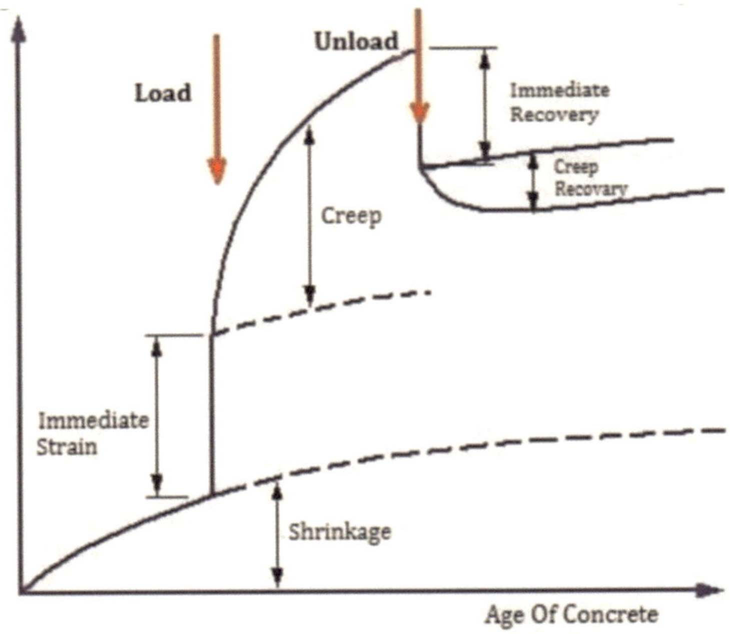

Figure 1.
Time dependent creep deformation.

strain and partly because of plastic strain or permanent deformation as shown in the **Figure 1**.

Creep testing is conducted on unsealed and sealed specimens with unsealed being the most used method of creep testing. These specimens, without an applied stress have volumetric changes due to drying and autogenous shrinkage. Hence, the total deformation of unsealed specimens is because of an applied stress producing elastic defamation, shrinkage and creep. Creep includes drying and basic creep; whereby, drying creep is the total deformation minus elastic deformation whereas basic creep is total deformation of loaded, sealed specimens minus elastic deformation and autogenous shrinkage. Autogenous shrinkage and basic creep require testing of both sealed and unsealed specimens [2].

2. Mechanism of creep

Creep of concrete is influenced by all its constituents as well as the loading time, duration of loading and environmental conditions. In general, the constituents used, age of loading, type and duration of loading mainly affect magnitude of creep. Environmental conditions affect not only amplitude but also the development of creep. Concrete behaves as a viscoelastic material with early age at loading and longer loading duration increasing creep.

Aggregates being stiffer than hydrated cement paste do not creep. Hence, a higher content of stiffer aggregates of notional size restrains concrete creep. Creep is also related to cement paste content though the relationship is not linear. Other relevant factors of concrete creep are; water-cement ratio, type of cement and its fineness, compressive strength, stress-strength ratio, environmental conditions, size of structural a member and admixtures and additives [3].

The long-term creep deformation mechanism in cement gel is that involving narrowing of intercrystallite spaces. On application of load, there is instantaneous elastic response from both the solid and liquid systems of the concrete matrix. Under sustained load the compressed liquid begins to migrate from higher to lower stressed areas. This is accompanied by the transfer of load from liquid to the surrounding solid. It is believed that after several days of sustained load pressure on capillary water disappears. As the process of hydration is progressing; growth of solid phase and the expense of liquid phase gradually changes the parameter governing extent and rate of creep [4].

3. Mathematical modelling of creep

The necessity of adapting any creep theory for the practical calculations of concrete constructions requires the presentation of the relationships between concrete deformations and time, in the form mathematical functions. These mathematical functions should be in good conformity with experimental results.

Presentation of creep functions should satisfy some determined conditions. The creep function denoted by C (t, τ) expresses creep deformation at any moment caused by unit stress at a time τ. These established conditions are given in Eqs. (1)-(7) as follows;

i. Creep at the time of loading is equal to zero for concrete of any age.

$$C(t,t) = C(t,\tau) \cong 0 \tag{1}$$

ii. For all time ‘t’ whereby t > τ; creep is more than 0.

$$C(t,\tau) > 0 \, for \, t > \tau \tag{2}$$

iii. The rate of increase of creep deformation at constant time loading diminishes with increase of ‘t’ and equals zero at ‘t’ tends to infinity.

$$\lim_{t \to \infty} \frac{\partial C(t,\tau)}{\partial t} = 0 \tag{3}$$

iv. The rate of decrease of creep after unloading diminishes with increasing ‘t’ and for $t = \infty$ the value is zero. However, the creep function diminishes at the same time.

v. The creep function diminishes uniformly with increasing concrete age.

$$\frac{\partial C(t,\tau)}{\partial \tau} < 0 \tag{4}$$

The most general creep theory of concrete developed by several authors assumes partial deformation reversibility. Its course it represented in the **Figure 2**; whereby the curve the unloading tends to a determined final value rather than zero.

The expression $\varphi(\tau)$ is introduced to the creep function, in the form of

$$C(t,\tau) = \varphi(\tau).f(t-\tau) \tag{5}$$

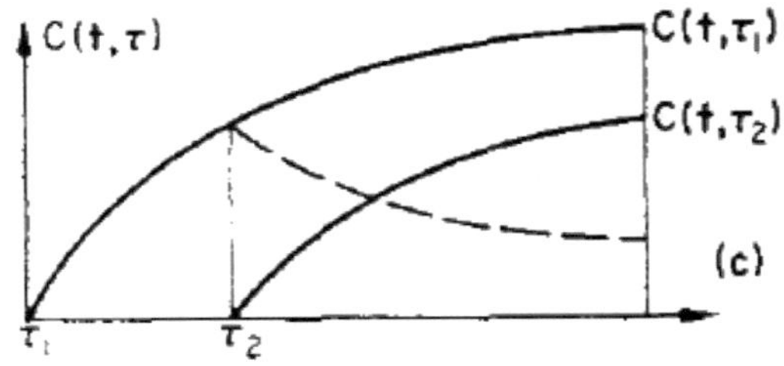

Figure 2.
General form of creep theory.

Ageing of concrete is also considered. The expression $f(t-\tau)$ is a loading duration function and allows partial reversibility deformations to be considered. Function $\varphi(\tau)$ accepts the form;

$$\varphi(\tau) = \left(C_0 + \frac{A_1}{\tau}\right) \tag{6}$$

Whereby C_0 is final value for $\tau \to \infty$

Finally, the creep theory function is of the form;

$$C(t,\tau) = \left(C_0 + \frac{A_1}{\tau}\right)\left[1 - e^{-y(t-\tau)}\right] \tag{7}$$

Whereby C_0, A_1 and y are determined in such a way that the theoretical and experimental curve match [5]. Creep functions lead to integral equations which after certain transformations can be reduced to easily solvable differential equations.

4. Creep prediction models

Various creep prediction models have been developed by researchers based on a large number of tests. The most commonly used models are; ACI-209 model developed by American Concrete institute. CEB-FIB model developed by Euro-international concrete committee. B3 model developed by Bazant Z. P and S. Baweja and the GL2000 model developed by Gardener and Lockman. All these models are fitting or empirical formulas from test data. These models look at free shrinkage, creep strain and elastic deformations. They relate creep strain to the loading conditions by using either creep compliance, specific creep or creep coefficient. Each model has its own reasonable explanation method based on theoretical evaluation and test data and therefore the explanation of their results has different formulas [6].

4.1 ACI model

Its procedure is applicable to normal and lightweight concrete (using types I and III cement, moist and steam-cured conditions) under standard conditions. This model is sensitive to water content.

Its mathematical formulations for elastic modulus, creep compliance and specific creep are shown in Eqs. (8)–(10) [7];

$$E_c(t) = 0.04326\sqrt{\rho^3 f_{cm}\left(\frac{t}{a+bt}\right)} \tag{8}$$

$$J(t,t_0) = \left[\frac{1+\Phi(t,t_0)}{E_c(t_0)}\right] \tag{9}$$

$$\Phi(t,t_0) = \frac{v_u(t-t_o)^{\varphi}}{d+(t-t_0)^{\varphi}} \tag{10}$$

4.2 CEB-FIB

Its prediction is restricted to ordinary structural concrete with 28 days mean cylinder strength ranging between 12 and 80 MPa, mean temperature of 5–30°C and mean relative humidity of 40–100%. This model is sensitive to relative humidity. The elastic modulus and creep compliance functions are given in Eqs. (11)-(15) as follows [7];

$$E_c(t_0) = E_c\sqrt{exp\left\{s\left(1-\frac{18}{t_o}\right)\right\}} \tag{11}$$

$$J(t,t_0) = \frac{1}{E_c(t_0)} + \frac{\Phi(t,t_0)}{E_c} \tag{12}$$

$$= \frac{1}{E_c(t_0)} + \frac{\Phi(t,t_0)}{E_c} \tag{13}$$

$$\Phi(t,t_0) = \left[1 + \frac{1-h}{0.46\left(\frac{2A_c}{100\mu}\right)^{\frac{1}{3}}}\right]\left[\frac{5.3}{\sqrt{0.1f_{cm}}}\right]\left[\frac{1}{0.1+(t_0)^{0.2}}\right] * \left[\frac{(t-t_0)^3}{\left[\beta_H+(t-t_0)^3\right]}\right] \tag{14}$$

$$\beta_H = 150\left[1+(1.2h)^{18}\right]\left(\frac{2A_c}{100\mu}\right) + 250 \le 1.50000 \tag{15}$$

4.3 B3 model

The prediction of material parameters for this model is limited to Portland cement concretes having w/c ratio of 0.30–0.85, a/c ratio of 2.5–13.5, cement content of 160–720 kg/m^3 and mean cylinder compressive strength varying between 17 and 70 MPa. Its mathematical formulations are given are sin Eqs. (16)-(26) as follows [7];

$$E_c(t) = 4.734\sqrt{\frac{f_{cm}t}{4+0.85t}} \tag{16}$$

$$J(t,t_0) = \frac{127}{\sqrt{f_{cm}}} + C_o(t,t_0) + C_d(t,t_0,t_c) \tag{17}$$

$$C_0(t,t_0) = 185.4c^{0.5}(f_{cm})^{-0.9}Q(t,t_0) + 53.766(w/c)^4 c^{0.5}(f_{cm})^{-0.9} \tag{18}$$

$$* \ln\{1+(t-t_0)^n\} + 20.3(a/c)^{-0.7}\ln(t,t_0) \tag{19}$$

$$Q(t,t_0) = \left[0.086(t_0)^{2/9} + 1.21(t_0)^{4/9}\right]^{-1} \tag{20}$$

$$* \left[+1 \left(\frac{\left[0.086(t_0)^{2/9} + 1.21(t_0)^{4/9}\right]^{-1}}{(t_0)^{-m} \ln\{1 + (t - t_0)^n\}} r(t_0) \right. \right]^{-1/r(t_0)} \tag{21}$$

$$r(t_0) = 1.7(t_0)^{0.12} + 8 \tag{22}$$

$$m = 0.5\, n = 0.1 \tag{23}$$

$$C_d(t,t_0,t_c) = 7.57 * 10^5 (f_{cm})^{-1 \in_{shu}{}^{-0.6}} \sqrt{e^{-8H(t)} - e^{-8H(t_0)}} \tag{24}$$

$$H(t) = [1 - (1 - h)s(t)]\ 0 \leq H(t) \leq 1 \tag{25}$$

$$\Phi(t,t_0) = E_c(t_0)J(t,t_0) - 1 \tag{26}$$

4.4 GL 2000 model

This model is applicable to concretes of w/c ratio 0.40–0.60 and characteristic compressive strength below 70 MPa. Its mathematical formulations are given in Eqs. (27)-(30) as follows [7];

$$E_c(t) = 3500 + 4300\sqrt{\left[\frac{f_{cm} t^{0.75}}{a_1 + a_2 t^{0.75}}\right]} \tag{27}$$

$$J(t,t_0) = \frac{1}{E_c(t_0)} + \textit{Specific creep} \tag{28}$$

$$\textit{Specific creep} = \frac{\Phi(t,t_0)}{E_c} \tag{29}$$

$$\Phi(t,t_0) = \Phi(t_c)\left[\left\{\frac{2.0(t - t_0)^{0.3}}{(t - t_0)^{0.3} + 1.4}\right\} + \sqrt{\left\{\frac{7(t - t_0}{t_0(t - t_0 + 7}\right\}} + 2.5(1 - 0.086h^2)\sqrt{\left\{\frac{t - t_0}{t - t_0} + 0.15(V/S)^2\right\}}\right] \tag{30}$$

The notations for all equations are given in Appendix E.

5. Review of creep of various concretes

5.1 Creep and shrinkage of concrete containing Iranian pozzolans

Ghodousi carried out a study on early -age creep and shrinkage of concrete containing Iranian Pozzolans. The pozzolans studies were silica fume (SFL), Trass (TL) and GGBS. These were compared with plain concrete (PL). ACI209, BS 8110-1986, CEB1970 prediction models and an estimation based on 28-day results were compared with experimental data. Silica fume concrete exhibited highest creep whereas plain concrete exhibited lowest creep for the 200 days of the test as shown the **Figure 3**.

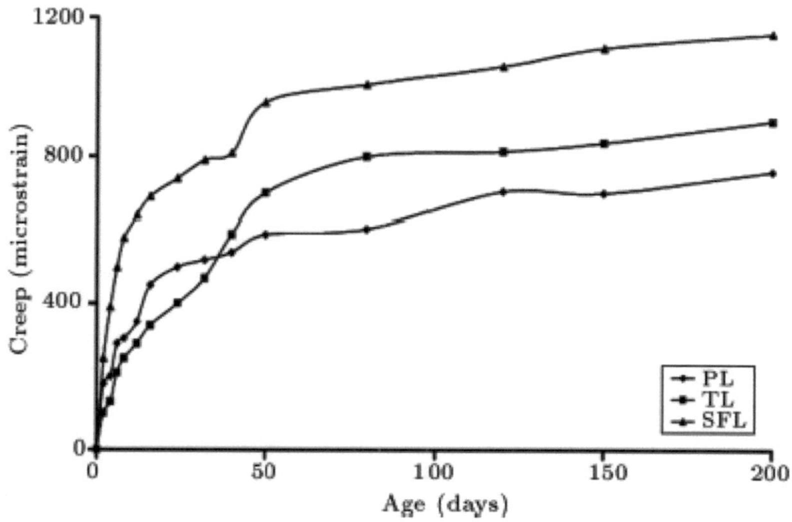

Figure 3.
200-days creep strain of concrete in laboratory conditions.

Concrete Code	ACI	BS	CEB	Estimates Based on 28-Day Results	Lab. Results
PL	564	930	846	972	760
TL	850	1399	1273	778	900
SFL	553	912	829	1421	1160
PC1	489	829	743	1777	1425
SFC1	492	829	743	1276	1695
SC1	759	1197	1086	1085	1000

Table 1.
200-day creep values based on tests and prediction.

In order to verify the accuracy of prediction models, predicted creep and shrinkage values as in **Table 1** were compared with measured values up to 200 days. It was found that most of the test results for creep and shrinkage strains were higher than predicted values. However, the 28-day estimation results (i.e., short-term data) showed a relatively better prediction of creep and shrinkage strains as shown in the table below. It was clearly seen that the accuracy of prediction of creep can be considerably improved by undertaking short-term test results of 28-days and extrapolating results to get long term values.

It was also observed that the type of pozzolanic material had major effects on creep and shrinkage. The large errors observed in predicting creep of concrete with pozzolanic materials point out that a more accurate model taking into account the effect of pozzolanic material content on the time function is required. Although the existing models include effects of many significant factors affecting creep, they have not taken into account the effect of pozzolanic materials in concrete. They concluded that for accurate predictions of creep, tests must be carried on prototype concrete and extrapolation for long term behaviour made [8].

5.2 Creep analysis of concrete containing rice husk ash

Zhi-hai carried out a study on compressive creep analysis of concrete containing rice husk ash experimentally. In the study the researcher varied the RHA content in concrete from 0 to 20%. The creep loading was 25% of the compressive strength of specimens at 28 days. Their results indicated that the specimen with higher RHA-to-binder ratio had a smaller creep strain as shown in **Figure 4**. However, the influence of RHA on creep was not linear as 10, 15 and 20% RHA in binder reduced the 60 days creep by 17, 30 and 33%. This indicated that 15% RHA was the best terms of creep reduction.

The phenomenon of decrease in creep with increasing RHA content was attributed to the pore microstructure of RHA. Large capillary pores exist in concrete without RHA whereas smaller capillary pores exist in concrete with RHA. Hence RHA decreases porosity in concrete improving durability properties of the material. In addition, RHA being highly reactive, reacts with CaOH originated from cement hydration producing additional C-S-H which fills larger pores or voids. Also the porous RHA particles release absorbed water retained in its small pores which helps improve cement hydration reducing the size of interfacial transition zones and porosity of gel pores [9]. Therefore, in cooperation of RHA in concrete makes the concrete matrix to be dense and impervious hence reducing its susceptibility to creep deformation.

5.3 Creep and drying shrinkage of concrete containing GGBS

A study on compressive creep and shrinkage of concrete containing varying amounts of GGBS (0, 20, 40 and 60%) was carried out. Creep recovery was also observed. The results as shown in the **Figure 5** indicated that although the shape of creep-time curves for all concrete was similar, there was an increase in creep with increase in the amount of GGBS. This was attributed to the free water available in the body of concrete. The chemical requirement of water is for cement hydration and a relatively smaller quantity for CaOH to react with GGBS. The second reaction being

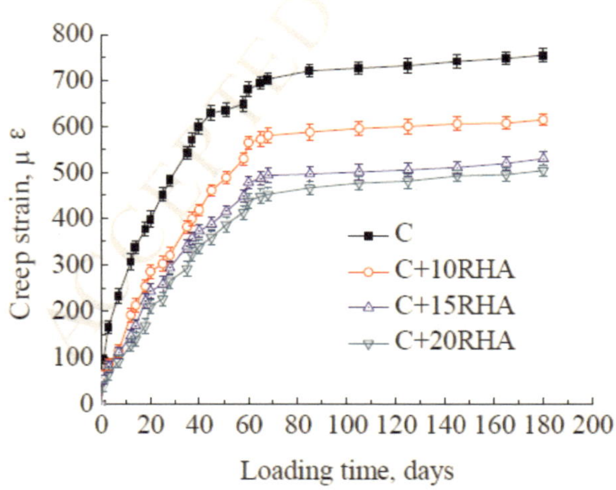

Figure 4.
Effect of RHA on creep strain of specimens.

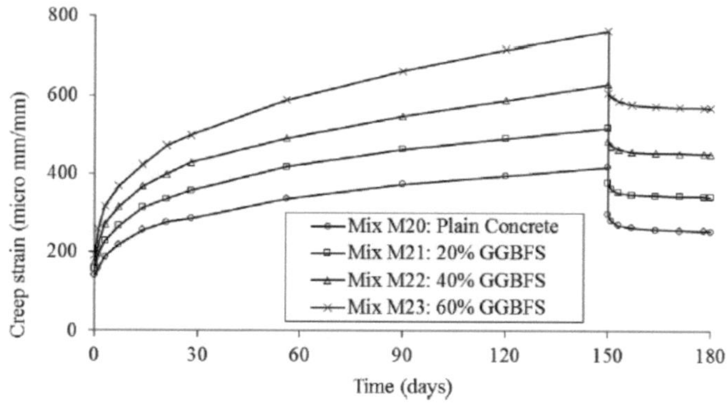

Figure 5.
Creep of GGBS concrete.

subsequent to the 1st leaves free water availability in GGBS being higher because a constant w/c is used for all mixes. This water gets squeezed out during loading. Therefore, a higher creep effect with increase in GGBS.

The range of parameters for creep models ACI, CEB, B3 and GL200 were not applicable for GGBS concrete. Therefore, a modification to incorporate the influence of GGBS was made. A new model was developed using multiple variable regression analysis if the test data. The creep coefficient was expressed as function of time, compressive strength and % of GGBS.

$$\Phi(t) = \frac{\alpha}{\sqrt{f_c^{\,\prime}}}\left(\frac{230t^{0.4})}{140 + 2.4t^{0.4})}\right. \tag{31}$$

$$\alpha = \exp(0.004p_s) \tag{32}$$

whereby
α is influence of GGBS on creep coefficient
t is time in days
$f_c^{\,'}$ is cylinder compressive strength of concrete
p_s is percentage of GGBS [10]

5.4 Experimental investigation of creep and shrinkage of R.C with influence of reinforcement ratio

Concrete creep and shrinkage with different ratios of reinforcements (0, 0.5, 1, 2 and 3.9%) was studied. Results of prediction by CEB, ACI, GL2000 and JTG D 62-2004 were compared with test data of plain concrete. The creep expressed by creep coefficient was found to increase faster at early age than latter age. The re-enforcement ratio was found to inhibit creep as shown in the **Figure 6** below. However low ratio of reinforcement was found to have little influence on creep.

Creep coefficients from the various models and the experiment showed a consistent trend with time. However, the prediction of ACI model was more accurate as shown in the **Figure** 7 below. Therefore, the ACI model was modified to fit the influence coefficient of reinforcement.

ACI model was reevaluated with test data of plain concrete giving the following equation for creep coefficient.

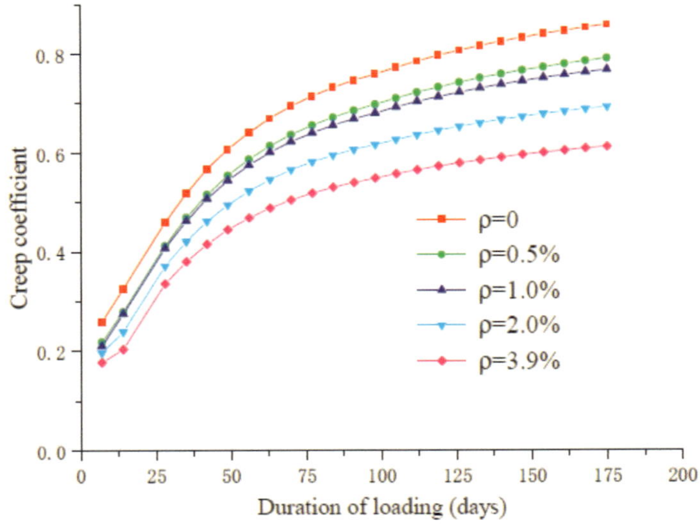

Figure 6.
Creep coefficient with varying reinforcement ratios.

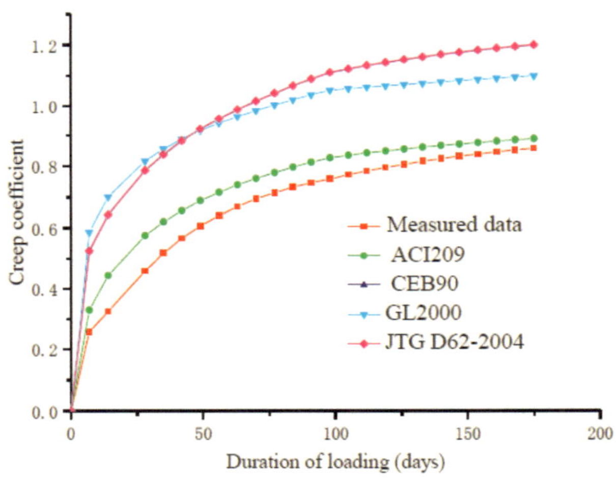

Figure 7.
Comparison of creep data with models.

$$\Phi(t, t_0 = 1.16x \frac{(t - t_o)^{0.95}}{25.87 + (t - t_o)^{0.95}} \tag{33}$$

Whereby t_0 is loading time.

t is time 1.16 is a factor related to composition of material and environment.

Then the influence coefficient of reinforcement was obtained by combining with test results for the different ratios of reinforcement to give the following equation.

$$\Phi(t, t_0) = 1.16x \frac{(t - t_0)^{0.95}}{25.87 + (t - t_0)^{0.95}} K_p \tag{34}$$

$$K_p = 0.83xe^{-0.072\rho}$$

Whereby K_p is reinforcement on creep influence
$\acute{\rho}$ is reinforcement ratio.

6. Conclusions

Creep in concrete has been defined and the mechanism which governs creep. Creep prediction models with the mathematical formulations has also been given. Lastly creep of some concrete has been explained and how the variations occur when constituents of concrete have been changed. It's recommended that more creep experiments be carried out with different types of concretes so as the increase the database of creep in concrete.

Conflict of interest

The author declares no conflict of interest.

Appendices and nomenclature

A_c	cross-sectional area
a, b	constants which depend on type on cement and curing
a_1, a_2	constants which depend on type of cement
a/c	aggregates cement ration by weight
C_d (t, t_o, t_c)	compliance function for simultaneous drying
C_o (t, t_o)	compliance function for basic creep (creep at constant moisture content)
c	cement content in concrete (kg/m^3)
d	constant normally between 6 and 30 days
d_{10}	10 days in absence of specific data for local aggregates and conditions
E_c	modulus of elasticity of concrete at 28 days (MPa)
$E_c(t)$	modulus of elasticity of concrete at age t (MPa)
$E_c(t_o)$	modulus of elasticity at age t_o (MPa)
f_{cm}	measured mean compressive strength at 28 days (MPa)
H(t)	spatial average of pore relative humidity within the cross section
h	relative humidity of environment at ambient temperature, decimal
J (t, to)	creep compliance representing the total stress dependent strain per unit stress
S	constant which depends on type of cement
S(t)	time dependence factor
t	age of concrete at time of observation(days)
t_c	age of concrete at which drying commenced
t_o	age of concrete at the time of loading (day)
V/S	volume to surface ratio(mm)
w	water content in concrete (kg/m3)
w/c	water-cement ratio by weight
β_{RH}	constants depend on relative humidity
ϵ_{shu}	notional ultimate shrinkage strain (10^{-6})
μ	perimeter of member in contact with atmosphere (mm)

v_u	coefficient of age at application of load
$\acute{\rho}$	unit weight of concrete (kg/m^3)
$\Phi(t, to)$	creep coefficient at time t
$\Phi(t_c)$	correction factor for drying before loading commenced
Ψ	constant normally between 0.40 and 0.80

Author details

Winfred Nthuka Mutungi
Jomo Kenyatta University of Agriculture and Technology, Nairobi, Kenya

*Address all correspondence to: mutungiwinfred@gmail.com

References

[1] Nicoara AI et al. End-of-life materials used as supplementary cementitious materials in the concrete industry. Materials (Basel). 2020;**13**(8):1-20

[2] Vincent EC. Compressive creep of a lightweight high-strength concrete mixture. Virginia Polytechnic Institute and State University MSc dissertation. 2003. p. 146

[3] Havlásek P. Creep and shrinkage of concrete subjected to variable environmental conditions. Czech Technical University in Prague Doctoral Dissertation. 2014. p. 204

[4] Shetty MS. Concrete Technology Theory and Practice Types of Cement. Vol. 0552000. New Delhi: S. Chand & Company Ltd. 2000. pp. 1-647

[5] Mitzel A. Concrete creep and shrinkage functions. Building Science. 1967;**2**(3):259-265

[6] Sun G, Xue S, Qu X, Zhao Y. Experimental investigation of creep and shrinkage of reinforced concrete with influence of reinforcement ratio. Advances in Concrete Construction. 2019;**7**(4):211-218

[7] Goel R, Kumar R, Paul DK. Comparative study of various creep and shrinkage prediction models for concrete. Journal of Materials in Civil Engineering. 2007;**19**(3):249-260

[8] Ghodousi P, Afshar MH, Ketabchi H, Rasa E. Study of early-age creep and shrinkage of concrete containing Iranian pozzolans: An experimental comparative study. Scientia Iranica. 2009;**16**(2 A): 126-137

[9] He Z h, Li L y, Du S g. Creep analysis of concrete containing rice husk ash. Cement and Concrete Composites. 2017;**80**:190-199

[10] Shariq M, Prasad J, Abbas H. Creep and drying shrinkage of concrete containing GGBFS. Cement and Concrete Composites. 2016;**68**:35-45

Chapter 5

Fire Resistance of Reinforced Concrete Slabs

Sanin Dzidic

Abstract

A fire can happen anytime and anywhere. That is why the fire resistance of our structures is important. Adequate fire resistance enables the evacuation of people and material goods, fire-fighting intervention, but also the possibility of reusing the building after the fire with minor or major retrofitting works. Our structures are made of different construction materials, and these materials behave differently in fire conditions, while we build our structures to last 50, 100, or 200 years. That is why the selection of structural and other construction materials is extremely important. Concrete is not combustible, does not transmit flame, does not release toxic gases in a fire and is not a fire load. However, fire has a negative impact on concrete structures as well, especially since concrete is a composite material, so there is a reduction in the mechanical properties of both concrete and the embedded steel. The most sensitive reinforced concrete elements in a fire situation are thin-walled elements such as RC slabs, which have a thin concrete cover. In this research, using four world famous methods, we try to figure out the factors influencing the fire resistance of RC slabs as the most sensitive reinforced concrete elements to fire.

Keywords: concrete, fire resistance, fire modeling, RC slabs, codes

1. Introduction

Despite modern firefighting methods and new technologies, fires still pose a great danger to both population and property. A fire is an uncontrolled burning of fuel that spreads uncontrollably in time and space and poses a danger to human life and material goods. We protect structures from fire using active and passive fire protection measures. One of the passive measures to protect structures from fire is the correct selection of both structural and other materials from which we build the structures and which we install into the building itself. Statistics in the developed countries of the world in the past 50 years show that the number of fires on structures is continuously decreasing, as well as the number of injured and dead people in those fires. This positive trend was significantly contributed by the increase in awareness of the dangers of fire, the development of science and technology in the field of fire protection, and the development and application of active and passive fire protection measures, but also the development of research focusing on the behavior of materials and structures in fire conditions, which resulted adopting modern standards. However, fire still remains a danger to people, property, and the economy.

In this regard, reaction to fire of different materials and fire resistance of different structures are of the most significant interest and consideration nowadays. The behavior of structures in fire conditions depending on the structural materials they are made of, size, shape and dimensions of the building, ventilation conditions, as well as thermal characteristics of the material applied at boundary surfaces of the observed building including applied fire protection active and passive measures, as well as type and density of fire load is significantly different. **Figure 1** shows the aftermath of one fire of the factory that was constructed in sections with masonry, steel, and concrete structures.

In addition, EU Regulation No. 305/2011 specifies "that construction works as a whole and in their separate parts must be fit for their intended use, taking into account in particular the health and safety of persons involved throughout the life cycle of the works" [1]. The same document sets up seven requirements that structures should fulfill, while the second one is related to safety in fire, immediately after the first one related to mechanical resistance and stability.

The same document also provides the following essential requirement for the limitation of fire risks: "The construction works must be designed and build in such a way, that in the event of an outbreak of fire, the load-bearing resistance of the construction can be assumed for a specified period of time, the generation and spread of fire and smoke within the works are limited, the spread of fire to neighboring construction works is limited, the occupants can leave the works or can be rescued by other means and the safety of rescue teams is taken into consideration" [1].

Among other reasons, this research is inline as a contribution to the aforementioned criteria, especially when it comes to concrete structures and reinforced concrete slabs as probably the most vulnerable concrete elements in fire conditions, wishing to contribute to a certain extent to both design and construction practice, but also in assessment of the existing structures that have experienced a fire and guidelines for possible retrofit of the such structures.

Figure 1.
Fire aftermath and structures made of different materials.

2. Research methodology

Qualitative comparative research methodology was used in this research. It aimed to determine fire resistance of the RC slabs using four methods. In the first stage, the available methods and their principles for determining fire resistance of concrete elements including main aspects of standard fires, properties of concrete and reinforcing steel at elevated temperatures, and fire resistance were discussed.

After that, empirical analysis of fire resistance of RC slabs of different spans, depth, concrete class, type of aggregate used, steel reinforcement, variation of concrete cover, and actions was done. This research was conducted to determine factors of influence that govern fire resistance of RC slabs.

There are three considered research questions in this study. The first one is related to the general fire resistance of RC slabs designed for ambient temperatures. The second is related to contribution of concrete class to the fire resistance of RC slabs, while the third research question focuses on the contribution of the thickness of concrete cover to fire resistance of RC slabs. The findings on these research questions will inform on the research hypothesis that proper selection of the concrete class and cover can significantly affect the fire resistance of RC slabs.

3. Modeling of fire as action to the structure

The most fires can be modeled through the curve of temperature and time dependence. How does a fire develop? At the very beginning, a fire in a room has the characteristics of an outdoor fire with a relatively low temperature and a limited range with a gradual increase in temperature. At this stage, the active fire protection measures are the most effective. If the active fire protection measures are not effective at this stage of fire development, the temperature in the room increases due to the preheating of the combustible material in the room, and the temperature rises to several hundred degrees. When the temperature reaches about 600°C in the higher zones of the room—ceiling, a complete flare-up of the fire occurs, which is in the literature called flashover (**Figure 2**).

This phenomenon looks spectacular, and the fire passes into the stage of a fully developed fire, when the action of active fire protection measures is limited, and fire very quickly reaches the maximum temperature.

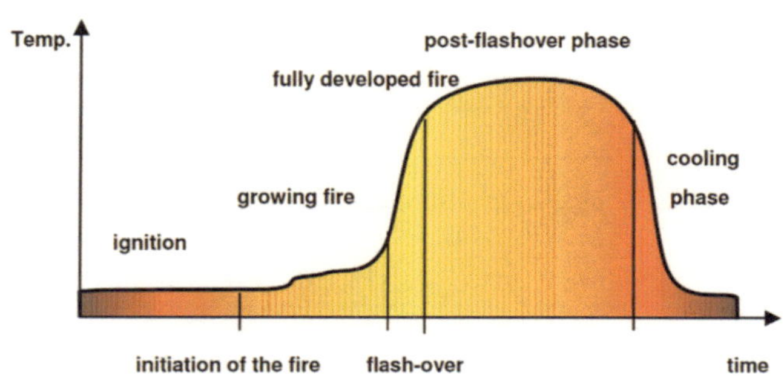

Figure 2.
Real fire model [2].

It is important to note that every real fire has an ascending and descending branch of fire development in the room. When the temperature in the room drops to 80% of the maximum developed temperature during the fire, it is considered that the burn-out phase has begun because the available amount of fuel that can release heat decreases, given that it has already been consumed in the previous phases of the fire.

Considering that every real fire develops in its own style, depending on many factors (e.g., shape, dimensions, room size, quantity, and type of fire load, and many other factors), it was necessary to define a standard fire to evaluate the fire resistance of structures, elements, and building products in fire conditions applying standardized procedures. The standard fire defined by The International Standard ISO 834-1—Fire Resistance Tests—Element of Building Construction from 1999 [3] is widely accepted for this purpose in most countries in the world. Eurocode 1 (EN 1991-1-2: 2002) [4] also adopts this temperature–time curve as a nominal curve. This standard fire curve is defined as:

$$T = 345\ log_{10}\ (8t + 1) + 20 \tag{1}$$

where.

T – average temperature in the test furnace in °C;

t – test time in minutes.

This standard fire curve is applied as action to structures for the determination of fire resistance of RC slabs for methods used for determining fire resistance of RC slabs in this research, namely EN 1992-1-2:2004, Eurocode 2, Design of concrete structures, Part 1–2: Structural fire design [5] and BRANZ Technical Recommendation No. 8 – Method for Fire Engineering Design of Structural Concrete Beams and Floor Systems [6].

The standard fire curve is used in the USA and along the ACI/TMS 216.1—Code Requirements for Determining Fire Resistance of Concrete and Masonry Construction Assemblies [7] is the standard fire curve temperature-time ASTM E 119 [8], and in this research it is presented with a number of discrete points as shown in **Table 1**.

Lie [9] gave several equations that mathematically approximate the ASTM E 119 curve, where the simplest one gives the temperature in function of time through the following relationship:

$$T = 750\left[1 - e^{-3.79553\sqrt{t_h}}\right] + 170.41\sqrt{t_h} + T_0 \tag{2}$$

Time (t) (min)	Temperature of standard fire ASTM E 119 (°C)
5	538
10	704
30	843
60	927
120	1010
240	1093
480 and over	1260

Table 1.
Temperature dependance for standard fire ASTM E 119 [8].

Figure 3.
Comparison of standard fire curves [10].

where

t_h – test time expressed in hours.

These two standard fire curves are rather similar but not the same, and a comparison is presented in **Figure 3**.

4. Methods for fire resistance determination of RC slabs

In order to approach the determination of the fire resistance of reinforced concrete slabs, the following considerations provide the basic principles of the available methods that are used in this research.

4.1 Eurocode 2: design of concrete structures, part 1-2: structural fire design

All Eurocodes relating to structural materials have a part 1–2 which deals with structural fire design, while Eurocode 1 has a part 1–2 related to actions on structures exposed to fire. Related to the EN 1992-1-2:2004, Eurocode 2, Design of concrete structures, part 1–2 [5], it deals with structural fire design of concrete structures. This European Standard defines three groups of methods for determining the fire resistance of concrete structures: tabular method, simplified calculation methods, and advanced calculation methods.

When determining the fire resistance of concrete structures or elements, nominal fire curves and parametric curves defined in Eurocode 1 can be used as a fire action to structure or element. The following three criteria are supposed to be considered in determining the fire resistance of concrete structures or elements:

- Mechanical resistance, load-bearing function (Criterion R);
- Integrity, separating function (Criterion E); and
- Insulation (Criterion I).

In accordance with EN 1992-1-2:2004—Structural Fire Design [5], Criterion "R" is considered fulfilled when the structure or its element retains its load-bearing function including meeting the stability conditions for the required fire exposure time. Integrity Criterion "E" represents the ability of the separating element to prevent the occurrence of fire and gases on the fire non-exposed side of the element, when exposed to standard fire on its other side and for the required period of time. The separating element meets the requirements of Insulation Criterion "I" if the average temperature rise on the side not exposed to fire does not exceed 140 K, and the maximal temperature rise at any point of the separating element does not exceed 180 K, when exposed to a standard fire on the other side, and during the required period of time.

Different elements should meet different criteria depending on their function in the building. Thus, a reinforced concrete slab or load-bearing reinforced concrete wall should meet all three criteria, while a reinforced concrete beam or column should only meet Criterion "R", and a non-load-bearing partition wall should meet criteria "E" and "I". [11].

Along the fire resistance analysis of the structural element, the reduction factor η_{fi} for actions in fire situation is applied in relation to the design at normal (ambient) temperature:

$$E_{d,fi} = \eta_{fi} \bullet E_d \tag{3}$$

where

$E_{d,fi}$—design effect of actions in the fire situation;

E_d—design effect of actions for the ambient temperature design;

η_{fi}—reduction factor for the design load level for the fire situation.

The values of partial safety factors in structural fire design for thermal and mechanical properties of materials (concrete and reinforcement) are $\gamma_{M,\,fi} = 1.0$.

4.1.1 Tabulated data

The tabulated data provide relatively simple and easy-applied solutions for everyday engineering practice and based on the "axial distance" of the reinforcement from the nearest fire-exposed surface of the concrete element when exposed to a standard fire up to 240 min for different types and functions of concrete elements. The data are given for concrete of normal weight (2000–2600 kg/m^3) made of quartz aggregate and can be also used for concrete made of calcareous aggregate applying a reduction of minimum dimensions of 10%. The axial distance "a" is defined as the distance from the centroid of the reinforcement to the nearest fire-exposed surface of the concrete element according to **Figure 4**.

The origin of data displayed in the tables for the minimum dimensions and axial distances of concrete elements to meet the Criterion "R" that comes from the following equation:

$$\frac{E_{d,fi}}{R_{d,fi}} \leq 1.0 \tag{4}$$

where

$E_{d,fi}$—design effect of actions in the fire situation; and.

$R_{d,fi}$—design load-bearing capacity in the fire situation.

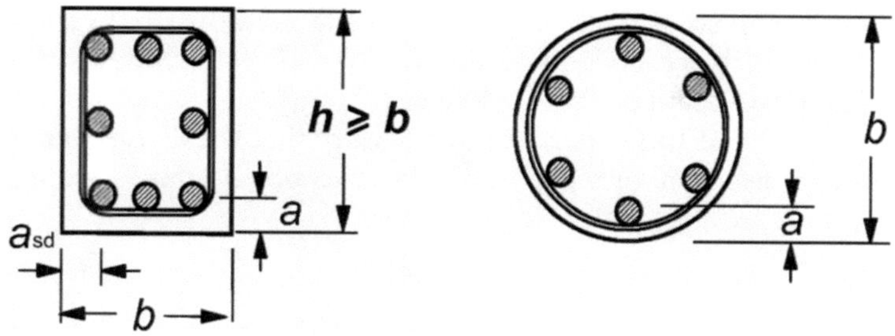

Figure 4.
Axis distance [5].

Standard fire resistance	Minimum dimensions (mm)			
	Slab thickness h_s (mm)	Axis distance a		
		One way	Two way	
			$l_y/l_x \leq 1.5$	$1,5 < l_y/l_x \leq 2$
1	2	3	4	5
REI 30	60	10*	10*	10*
REI 60	80	20	10*	15*
REI 90	100	30	15*	20
REI 120	120	40	20	25
REI 180	150	55	30	40
REI 240	175	65	40	50

**Normally, the cover required by EN 1992-1-1 will control.*
l_x and l_y are the spans of a two-way slab (two directions at right angles) where l_y is the longer span.
For prestressed slabs, the increase of axis distance according to 5.2(5) should be noted.
The axis distance a in Column 4 and 5 for two-way slabs related to slabs supported at all four edges. Otherwise, they should be treated as one-way spanning slab.

Table 2.
Minimum dimensions and axis distance for simply supported concrete slabs [5].

Axis distance given in **Table 2**, Column 3, for one-way simply supported slabs are based on critical temperature of the steel reinforcement Θ_{cr} = 500°C and the reduction factor for design load level in the fire situation η_{fi} = 0.7, which corresponds to $E_{d,fi}$ = 0.7 E_d, where E_d is designed effect of action at normal (ambient) temperature according to EN 1992-1-1 and partial safety factor for steel γ_s = 1,15 at ambient temperature. For satisfying the Criteria E and I, the minimal depth of slab is also prescribed in **Table 2**, Column 2.

4.1.2 Simplified calculation method for beams and slabs

The Eurocode 2, part 1–2, Annex E [5] provides a simplified calculation procedure for a more precise determination of the fire resistance of concrete beams and slabs when the actual axial distance of the centroid of the reinforcement from the nearest surface exposed to fire has a smaller value than prescribed in the tabular method, or a more precise extension of tabular method. It refers to beams and slabs designed by methods of linear analysis or linear analysis with a limited distribution at ambient temperature, and which are loaded predominantly with uniformly distributed load along the length of the beam or slab surface.

The main guideline of this procedure is to ensure that the design bending moment in fire conditions is less than or equal to the cross-sectional moment capacity in the fire situation:

$$M_{Ed,fi} \leq M_{Rd,fi} \tag{5}$$

For the simply supported slab, the cross-sectional moment capacity in the fire situation is determined according to the following relationship:

$$M_{Rd,fi} = \left(\frac{\gamma_s}{\gamma_{s,fi}}\right) \bullet k_s\,(\theta) \bullet M_{Ed} \bullet \left(\frac{A_{s,prov}}{A_{s,req}}\right) \tag{6}$$

where.

γ_s—the partial safety factor for steel used for ambient temperature.

$\gamma_{s,fi}$—the partial safety factor for steel in fire conditions;

$k_s\,(\theta)$—steel strength reduction factor for the given temperature θ for the required time of fire resistance;

M_{Ed}—designed bending moment for ambient temperature design;

$A_{s,prov}$—actual cross-sectional area of tensile steel reinforcement provided; and.

$A_{s,req}$—design required cross-sectional area of tensile steel reinforcement at the ambient temperature according to EN 1992-1-1 [12].

It is important to note that ratio of actual cross-sectional area of tensile steel reinforcement provided $A_{s,\ prov}$, and design required cross-sectional area of tensile steel reinforcement at the ambient temperature $A_{s,\ prov}$ should not be greater than 1.3.

Temperature of concrete in fire situation θ can be determined based on temperature profiles for slabs given in EN 1992-1-2: 2004 Annex A, [5], as shown in **Figure 5**.

In determination of fire resistance of concrete elements, it is common to assume that temperature of concrete and temperature in the reinforcement are the same at the same fiber of concrete element in fire conditions. Using this assumption, the steel strength reduction factor $k_s(\theta)$ at the temperature θ for the required fire resistance period can be determined from the next diagram (**Figure 6**).

4.2 BRANZ technical recommendation No. 8: method for fire engineering design of structural concrete beams and floor systems

Based on the results of a large number of tests on the fire resistance of concrete elements, which were sponsored by the Portland Cement Association in the USA, a procedure for determining the fire resistance of reinforced concrete and prestressed beams and slabs was developed by BRANZ – The Building Research Association of New Zealand and was published as BRANZ Technical Recommendation No. 8 [6]. The specific of this method is that represents a combination of principles for determining fire resistance according to ACI 216.1–97/TMS 0216.1–97, but in this procedure the action on the structure can also be modeled according to standard fire ISO 834-1.

Among other specifics, the method is applicable for determining the fire resistance of the simply supported or continious concrete slabs with or without restraints against

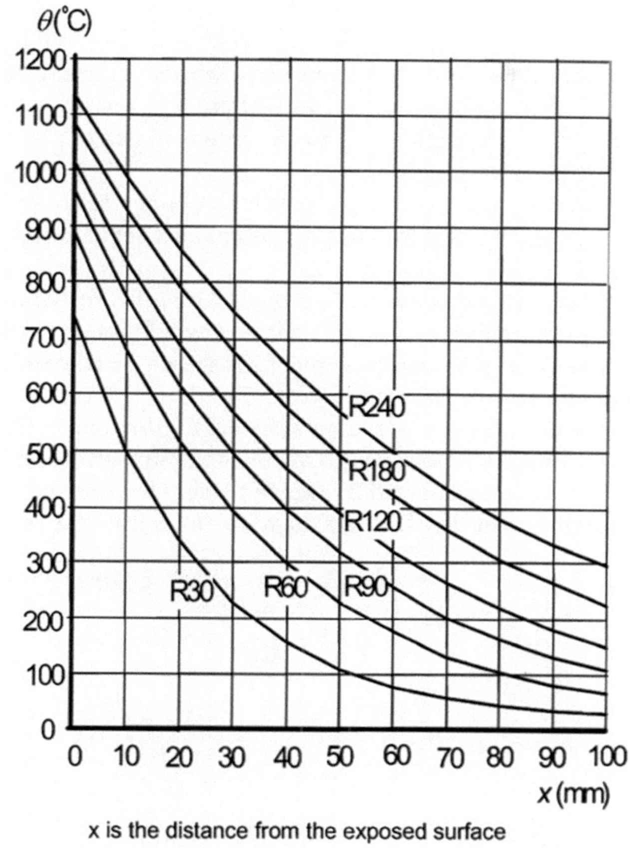

Figure 5.
Temperature profiles for slabs (depth 200 mm) [5].

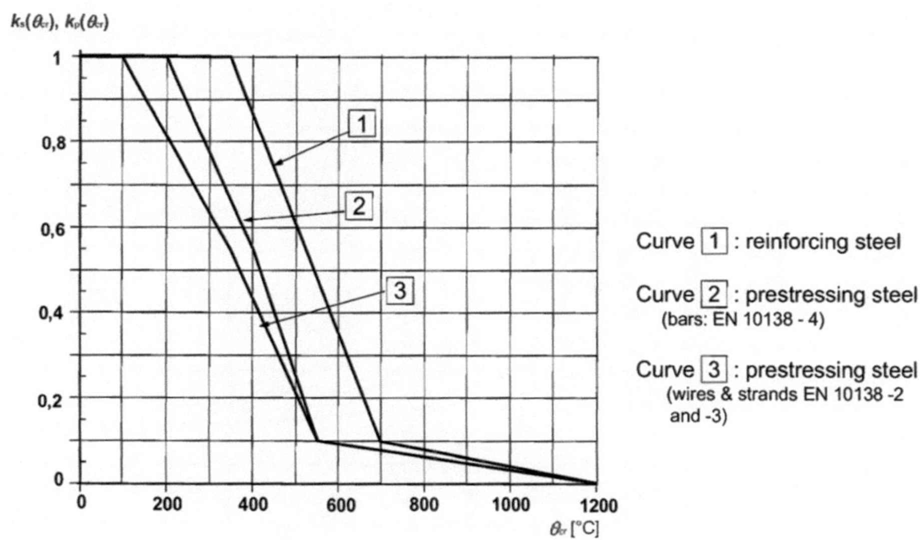

Figure 6.
Steel strength reduction factors [5].

termal expansion, made of normal weight or lightweight concrete as reinforced or prestressed. It is possible to determine the resistance to fire from 60 to 240 minutes.

The simply supported reinforced concrete slabs without restraints against thermal expansion and made of concrete of normal weight are focus of this research, so the aspects of this method related exclusively to the stated subject of research are presented in the further considerations.

The procedure begins by determining the effective concrete cover C_e as the shortest distance between the centroid of the individual steel bars and the fire-exposed surface of the slab.

Based on a series of fire resistance tests in the USA by the Portland Cement Association, the diagram in **Figure 7** was developed, which is used to determine the nominal temperature of the steel reinforcement for the required period of fire resistance as a function of effective concrete cover.

Once the nominal temperature of the tensile steel reinforcement T_s is determined, it is possible to determine its yield strength at elevated temperature using equation (Eq. (7)), or via the diagram displayed in **Figure 8** using the bilinear diagram "a" for the concrete reinforcing steel. The bilinear diagram "b" is used for prestressing steel:

$$\frac{f_y\,(T)}{f_y\,(20°C)} = f(x) = \begin{cases} 1.0 & for\ T \leq 250°C \\ 1.53 - \dfrac{T}{470} & for\ T > 250°C \end{cases} \qquad (7)$$

where

T—temperature of steel in °C;

f_y (20°C)—characteristic yield strength of reinforcing steel at normal-ambient temperature;

f_y (T)—yield strength of reinforcing steel at temperature T in °C.

In order to satisfy the equilibrium conditions in cross section of the slab, it is necessary to know the dimensions of the effective compression block of concrete and

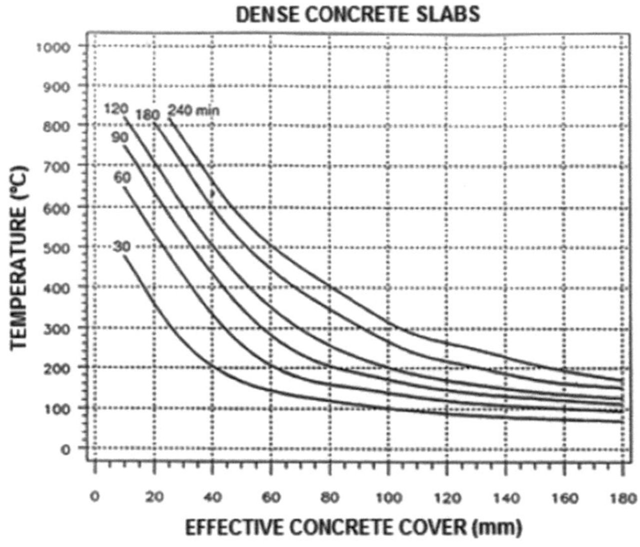

Figure 7.
Reinforcing steel temperature determination [6].

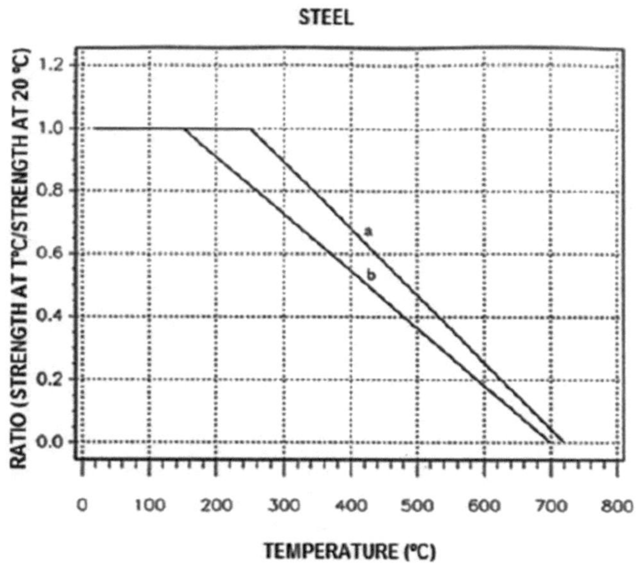

Figure 8.
Ratio of steel yield strength at elevated temperatures [6].

the compressive strength of concrete in fire conditions. According to this method, the effective compression concrete in fire conditions is considered when concrete temperature does not exceed 750°C. However, considering that the concrete compression block of simply supported concrete slab is not directly exposed to fire, the width of the compression concrete blok is equal to 1.0 m, and in most of these cases the temperature of the compression concrete block is below 350°C. Then the depth of the compressiion concrete block (a_Θ) can be determined according to the following formula:

$$a_\theta = \frac{A_s \bullet f_{y\theta}}{0.85 \bullet f'_{c\theta} \bullet b_\theta} \tag{8}$$

where

a_θ—depth of the equivalent rectangular compression stress block at the required time of the fire resistance;

A_s—cross-sectional area of tensile reinforcing steel provided;

$f_{y\theta}$—yield stress of reinforcement at the temperature θ;

$f'_{c\theta}$—reduced compressive strength of concrete at the temperature θ;

b_θ—reduced width of concrete compression block at the temperature θ (equal 1.0 m for slabs).

For the determination of compressive strength of normal-weight concrete at elevated temperatures, the following equation is used:

$$\frac{f'_c(T)}{f'_c\,(20°C)} = \begin{cases} 1.0 & \textit{for } T \le 350°C \\ 1.65 - 0.8 \bullet \dfrac{T}{440} & \textit{for } T > 350°C \end{cases} \tag{9}$$

where

T—temperature of compressive concrete in °C;

f'_c (20°C)—characteristic compression strength of concrete at the normal-ambient temperature;

f'_c (T)—compression strength of concrete at the temperature T.

As noted earlier, in most cases, the temperature of concrete in compression is below 350°C, so the compression strength of concrete in a fire situation is equal to the compression strength of concrete at normal-ambient temperature. This fact simplifies the procedure for concrete slabs.

Then, the bending moment capacity of the slab M_θ^+ in fire situation is as follows:

$$M_\theta^+ = A_s \bullet f_{y\theta} \bullet \left(d_\theta - \frac{a_\theta}{2}\right) \tag{10}$$

where

d_θ—effective depth of the slab in fire situation.

The safety factor for permanent actions in fire situation is 1.0 according to this Technical Recommendation, while the safety factor for the variable actions depends on the type of structure. For domestic, office, parking, and trafficable roofs, the safety factor for variable actions is 0.4, for floors in storages and other structures is 0.6, while it is 0 for non-trafficable roofs. The slab has the required fire resistance if the bending moment capacity is greater than the maximum applied bending moment in a fire situation.

4.3 ACI/TMS 216.1—code requirements for determining fire resistance of concrete and masonry construction

Based on research conducted by Martin, Gustaffero, Bletzacker, Harmathy, Abrams, and others, ACI/TMS Committee 216 developed a standard method for determining fire resistance of concrete and masonry structures presented in ACI/TMS 216.1 [7].

This method enables the design of fire-resistant concrete structures as well as the assessment of the fire resistance of existing concrete structures when exposed to standard fire ASTM E-119. The combination of actions in fire situation used in this method is prescribed in Appendix C2.5 from ASCE 07, Minimum Design Loads for Buildings and other Structures [8] for the calculation of moment due to full-service load on member M.

On the basis of structural end-point behavior, the fire resistance of a simply supported, unrestrained slab is as follows:

$$M_n \geq M_{n\theta} \geq M \tag{11}$$

where

M_n—the nominal moment capacity at section;

$M_{n\theta}$—the nominal moment capacity of section at elevated temperature;

M—the moment due to full-service load on a member.

If the reinforcement is straight and equal along the entire span of the slab, then the nominal moment capacity at section is also equal along the entire span of the slab, and it is as follows:

$$M_n = A_s \bullet f_y \bullet \left(d - \frac{a}{2}\right) \tag{12}$$

where

A_s—cross-sectional area of longitudinal tension reinforcement;

f_y—specified yield strength of reinforcing steel;

d—effective depth, distance from centroid of tension reinforcement to extreme compressive fiber;

a—depth of equivalent rectangular concrete compressive stress block at nominal flexural strength.

$$a = A_s \bullet f_y \bullet \left(0.85 \bullet f'_c \bullet b\right) \tag{13}$$

where

f'_c—specified compressive strength of concrete;

b—1.0 m for slabs (width of compression zone).

It is generally assumed that dead and live loads remain constant during the fire. However, the strength of materials is reduced so that the nominal moment capacity of section at elevated temperature is

$$M_{n\theta} = A_s \bullet f_{y\theta} \bullet \left(d - \frac{a_\theta}{2}\right) \tag{14}$$

in which Θ signifies effects of elevated temperatures. Note that cross-sectional area of longitudinal tension reinforcement A_s and effective depth d are not affected but yield strength of reinforcing steel $f_{y\theta}$ is reduced. Similarly, depth of equivalent rectangular concrete compressive stress block a_θ is reduced, but the compressive strength

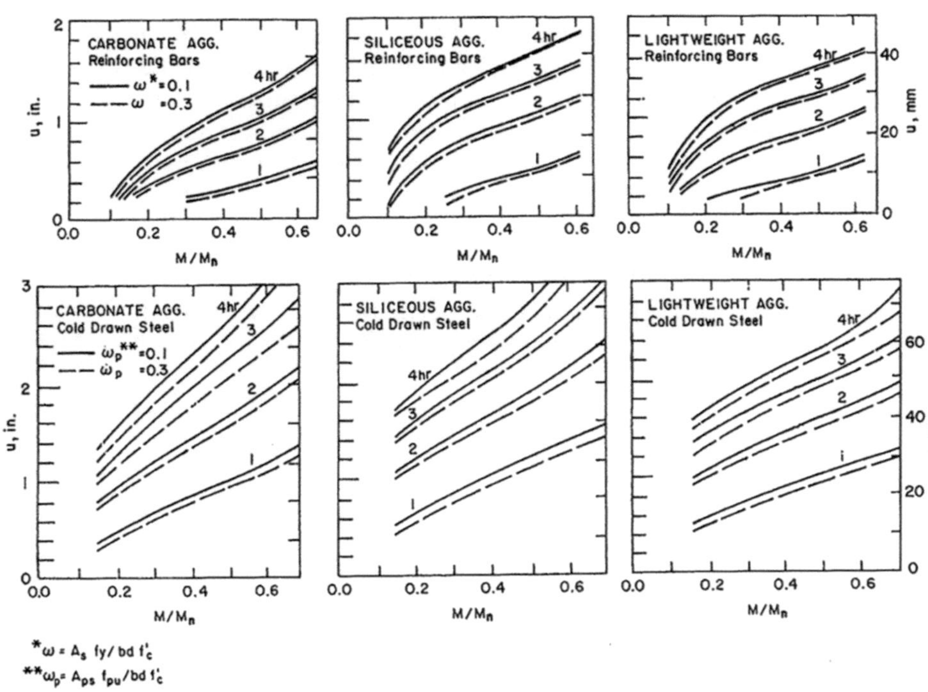

Figure 9.
Fire resistance of concrete slabs as influenced by aggregate type, reinforcing steel type, moment intensity, and "u" [7].

of concrete strength at the top of the slab f'_c is generally not significantly reduced. In addition, the period of fire resistance of the observed slab depends on the time required to reach the critical temperature of the steel, which again depends on the applied concrete cover.

Depending on the type of aggregate from which the concrete is made, the type of reinforcement, the ratio of moment due to full-service load and the nominal moment capacity M/M_n, and the average thickness of concrete between the center of main reinforcing steel and fire-exposed surface u, calculating the reinforcement index for concrete slab as:

$$\omega = \frac{A_s \bullet f_y}{b \bullet d \bullet f'_c} \tag{15}$$

it is possible to determine fire resistance of slab using diagrams given in **Figure 9**.

5. Determination of RC slabs fire resistance

In this research, we consider the fire resistance of simply supported reinforced concrete slabs, spans of 3, 5, and 7 m, and slab thicknesses of 12, 15, and 17 cm, respectively. For each span and slab height, the concrete classes C20/25, C 30/37, and C 40/50 are varied, wishing to establish the influence of different concrete classes on the fire resistance of RC slabs. The thickness of the concrete cover also varies from 0.5 to 3.0 cm, in increments of 0.5 cm. The use of thicker concrete covers in reinforced concrete slabs is not rational, so they are not used in this research. We are aware that EN 1991-1-1: 2004 does not allow the use of concrete cover of 0.5 and 1.0 cm, but the reason for the analysis of these concrete cover thicknesses enables to more precisely establish the dependence and contribution of concrete cover to the fire protection of reinforcement and reaching its critical temperature, while with on the other hand, it is neither impossible nor a rare, that during the execution of works, the thickness of the concrete cover is realistically smaller than designed in the actual

Fire resistance of slab Depth 12 cm, span 3 m, C 20/25, q_d = 9.075 kN/m^2								
Concrete cover (cm)	**a_s (cm)**	**$M_{Ed, fi}$ (kNm/m')**	**Reinforcement**	**Bar diameter (mm)**	**EN 1992-1-2 simplified method for slabs**	**EN 1992-1-2 tabulated data**	**BRANZ TR 8**	**ACI/ TMS 216.1**
0.5	0.825	7.14	R 221	6.5	R Ø	R Ø	0 min	<60 min
1.0	1.275		R 238	5.5	R 30	R 30	30 min	<60 min
1.5	1.850		R 257	7.0	R 30	R 30	30 min	<60 min
2.0	2.350		R 257	7.0	R 60	R 60	60 min	<60 min
2.5	2.800		R 283	6.0	R 60	R 60	60 min	60 min
3.0	3.325		R 335	6.5	R 90	R 90	90 min	60 min

Table 3.
Fire resistance of RC slab, span 3 m, depth 12 cm, C 20/25.

construction. All slabs are reinforced with welded reinforcing mesh of steel grade B500A, Ductility Class A, Yield = Re 500 MPa or individual rebar of steel grades B500A or St-500-b.

The slabs are previously designed according to EN 1991-1-1: 2004 at ambient-normal temperature. Thereby, the self-weight of the slab and the layers of the floor structure of 1.5 kN/m^2 are assumed as permanent actions, while variable load of 2 kN/m^2 was taken in design. For all variations of concrete slabs, fire resistance was determined according to all four methods described in this research. The fire was modeled using standard fire according to ISO 834-1 for the first three methods, while standard fire ASTM E 119 was used for the fourth method. The results of this research are shown in **Tables 3–11**.

Fire resistance of slab Depth 12 cm, span 3 m, C 30/37, q_d = 9.075 kN/m^2								
Concrete cover (cm)	**a_s (cm)**	**$M_{Ed, fi}$ (kNm/m')**	**Reinforcement**	**Bar diameter (mm)**	**EN 1992-1-2 simplified method for slabs**	**EN 1992-1-2 tabulated data**	**BRANZ TR 8**	**ACI/TMS 216.1**
0.5	0.825	7.14	R 221	6.5	R Ø	R Ø	0 min	< 60 min
1.0	1.300		R 238	5.5	R 30	R 30	30 min	< 60 min
1.5	1.850		R 257	7.0	R 30	R 30	30 min	< 60 min
2.0	2.350		R 257	7.0	R 60	R 60	60 min	< 60 min
2.5	2.800		R 283	6.0	R 60	R 60	60 min	60 min
3.0	3.325		R 335	6.5	R 90	R 90	90 min	60 min

Table 4.
Fire resistance of RC slab, span 3 m, depth 12 cm, C 30/37.

Fire resistance of slab Depth 12 cm, span 3 m, C 40/50, q_d = 9.075 kN/m^2								
Concrete cover (cm)	**a_s (cm)**	**$M_{Ed, fi}$ (kNm/m')**	**Reinforcement**	**Bar diameter (mm)**	**EN 1992-1-2 simplified method for slabs**	**EN 1992-1-2 tabulated data**	**BRANZ TR 8**	**ACI/TMS 216.1**
0.5	0.825	7.14	R 221	6.5	R Ø	R Ø	0 min	<60 min
1.0	1.275		R 238	5.5	R 30	R 30	30 min	<60 min
1.5	1.775		R 238	5.5	R 60	R 30	30 min	<60 min
2.0	2.350		R 257	7.0	R 60	R 60	60 min	<60 min
2.5	2.800		R 283	6.0	R 90	R 60	60 min	60 min
3.0	3.300		R 283	6.0	R 90	R 90	60 min	60 min

Table 5.
Fire resistance of RC slab, span 3 m, depth 12 cm, C 40/50.

FIRE RESISTANCE OF SLAB Depth 15 cm, span 5 m, C 20/25, q_d = 10.08 kN/m²								
Concrete cover (cm)	a_s (cm)	$M_{Ed, fi}$ (kNm/m')	Reinforcement	Bar diameter (mm)	EN 1992-1-2 simplified method for slabs	EN 1992-1-2 tabulated data	BRANZ TR 8	ACI/ TMS 216.1
0.5	0.95	22.03	R 636	9	R 30	R Ø	0 min	<60 min
1.0	1.45		R 636	9	R 30	R 30	30 min	<60 min
1.5	1.95		R 636	9	R 60	R 30	30 min	<60 min
2.0	2.45		R 636	9	R 60	R 60	60 min	<60 min
2.5	3.00		R 785	10	R 60	R 90	60 min	60 min
3.0	3.50		R 785	10	R 90	R 90	90 min	60 min

Table 6.
Fire resistance of RC slab, span 5 m, depth 15 cm, C 20/25.

Fire resistance of slab Depth 15 cm, span 5 m, C 30/37, q_d = 10.08 kN/m²								
Concrete cover (cm)	a_s (cm)	$M_{Ed, fi}$ (kNm/m')	Reinforcement	Bar diameter (mm)	EN 1992-1-2 simplified method for slabs	EN 1992-1-2 tabulated data	BRANZ TR 8	ACI/ TMS 216.1
0.5	0.95	22.03	R 636	9	R 30	R Ø	30 min	<60 min
1.0	1.45		R 636	9	R 30	R 30	30 min	<60 min
1.5	1.95		R 636	9	R 60	R 60	30 min	<60 min
2.0	2.50		R 785	10	R 90	R 90	90 min	60 min
2.5	3.00		R 785	10	R 90	R 90	90 min	60 min
3.0	3.50		R 785	10	R 120	R 90	90 min	60 min

Table 7.
Fire resistance of RC slab, span 5 m, depth 15 cm, C 30/37.

6. Conclusion

This parallel comparison research for determining fire resistance of reinforced concrete slabs of different spans and depths, made of three different concrete classes, with variations of concrete covers ranging from 0.5 m to 3.0 cm (with increments of 0.5 cm), applying four different procedures provided the following conclusions:

- The fire resistance periods of reinforced concrete slabs determined by four different methods are similar but not identical. The maximum difference in fire resistance periods of reinforced concrete slabs, depending on the method applied, is 60 min;

- Both methods used in this study specified in EN 1992-1-2:2004 for determining the fire resistance of RC slabs give identical or very similar fire resistance periods.

Fire resistance of slab Depth 15 cm, span 5 m, C 40/50, q_d = 10.08 kN/m^2								
Concrete cover (cm)	**a_s (cm)**	**$M_{Ed, fi}$ (kNm/m')**	**Reinforcement**	**Bar diameter (mm)**	**EN 1992-1-2 simplified method for slabs**	**EN 1992-1-2 tabulated data**	**BRANZ TR 8**	**ACI/ TMS 216.1**
0.5	0.95	22.03	R 636	9	R 30	R ∅	30 min	<60 min
1.0	1.45		R 636	9	R 30	R 30	30 min	<60 min
1.5	1.95		R 636	9	R 60	R 30	30 min	<60 min
2.0	2.50		R 636	9	R 60	R 60	60 min	60 min
2.5	3.00		R 785	10	R 90	R 90	60 min	60 min
3.0	3.50		R 785	10	R 120	R 90	90 min	60 min

Table 8.
Fire resistance of RC slab, span 5 m, depth 15 cm, C 40/50.

Fire resistance of slab Depth 17 cm, span 7 m, C 20/25, q_d = 10.76 kN/m^2								
Concrete cover (cm)	**a_s (cm)**	**$M_{Ed, fi}$ (kNm/m')**	**Reinforcement**	**Bar diameter (mm)**	**EN 1992-1-2 simplified method for slabs**	**EN 1992-1-2 tabulated data**	**BRANZ TR 8**	**ACI/ TMS 216.1**
0.5	1.10	46.12	R 1130	12	R 30	R 30	30 min	<60 min
1.0	1.60		R 1130	12	R 30	R 30	30 min	<60 min
1.5	2.20		∅ 14/12.5 cm	14	R 60	R 60	60 min	<60 min
2.0	2.70		∅ 14/12.5 cm	14	R 90	R 60	60 min	<60 min
2.5	3.20		∅ 14/10 cm	14	R 120	R 90	90 min	60 min
3.0	3.70		∅ 14/10 cm	14	R 120	R 90	90 min	60 min

Table 9.
Fire resistance of RC slab, span 7 m, depth 17 cm, C 20/25.

This fact is expected, considering that the Simplified Method is an extension of the Tabular Method for more precise determination of the period of fire resistance. The maximum difference between these two methods in periods of resistance to fire is 30 min;

- The fire resistances of RC slabs determined according to BRANZ TR 8 in the vast majority of cases have the same values as the methods specified in Eurocode 2. When differences exist, BRANZ TR 8 is more conservative;

- The ACI/TMS 216.1 method is not sensitive for the determination of fire resistance periods of RC slabs below 60 min. The results obtained using this method are not fully comparable with the other three methods considered in this study, given that the action in fire conditions is modeled in accordance with the standard fire ASTM E 119, while for the other three methods the standard fire

FIRE RESISTANCE OF SLAB Depth 17 cm, span 7 m, C 30/37, q_d = 10.76 kN/m^2								
Concrete cover (cm)	**a_s (cm)**	**$M_{Ed,\,fi}$ (kNm/m')**	**Reinforcement**	**Bar diameter (mm)**	**EN 1992-1-2 simplified method for slabs**	**EN 1992-1-2 tabulated data**	**BRANZ TR 8**	**ACI/ TMS 216.1**
0.5	1.10	46.12	R 1130	12	R 30	R 30	30 min	<60 min
1.0	1.60		R 1130	12	R 30	R 30	30 min	<60 min
1.5	2.20		∅ 14/12.5 cm	14	R 60	R 60	30 min	<60 min
2.0	2.70		∅ 14/12.5 cm	14	R 90	R 60	60 min	60 min
2.5	3.20		∅ 14/12.5 cm	14	R 90	R 90	60 min	60 min
3.0	3.70		∅ 14/10 cm	14	R 120	R 90	90 min	60 min

Table 10.
Fire resistance of RC slab, span 7 m, depth 17 cm, C 30/37.

Fire resistance of slab Depth 17 cm, span 7 m, C 40/50, q_d = 10.76 kN/m^2								
Concrete cover (cm)	**a_s (cm)**	**$M_{Ed,\,fi}$ (kNm/m')**	**Reinforcement**	**Bar diameter (mm)**	**EN 1992-1-2 simplified method for slabs**	**EN 1992-1-2 tabulated data**	**BRANZ TR 8**	**ACI/ TMS 216.1**
0.5	1.10	46.12	R 1130	12	R 30	R 30	30 min	<60 min
1.0	1.60		R 1130	12	R 30	R 30	30 min	<60 min
1.5	2.10		R 1130	12	R 60	R 60	30 min	<60 min
2.0	2.60		R 1130	12	R 60	R 60	60 min	<60 min
2.5	3.20		∅ 14/12.5 cm	14	R 90	R 90	60 min	60 min
3.0	3.70		∅ 14/10 cm	14	R 120	R 120	90 min	60 min

Table 11.
Fire resistance of RC slab, span 7 m, depth 17 cm, C 40/50.

ISO 834-1 is used. However, the development of temperatures depending on the fire exposure time in these two standard fires is not substantially significant, as can be noticed from **Figure 3**. Since the term—resistance to fire—in its broader meaning is not tied to the country, boundaries, or standards, this method was also used in this research. The differences observed in the fire resistance periods of RC slabs from this study determined by ACI/TMS 216.1 are up to 60 min if compared to the other three methods from this research;

- This research showed that the thickness of the concrete cover has a major influence on the fire resistance of RC slabs. The increased thickness of the concrete cover makes a difference in the fire resistance of RC slabs up to 90 min, provided that the concrete cover maintains its integrity without surface spalling of the concrete cover protecting the steel reinforcement. To further increase the fire resistance of RC slabs, if it is required, it is not rational to increase the thickness of the concrete cover, but to consider and investigate possible other methods for increasing the fire resistance of RC slabs;

- The maximum fire resistance of RC slabs observed in this research was 120 min.

Author details

Sanin Dzidic
Faculty of Technical Sciences, Department of Civil Engineering, University of Bihac, Bihac, Bosnia and Herzegovina

*Address all correspondence to: ninsa_d@hotmail.com

References

[1] The European Parliament and of the Council. Regulation (EU) No 305/2011 of 9 March 2011. 2011

[2] Lehner S. European Fire Classification of Construction Products, New Test Method "SBI", and Introduction of the European Classification System into German Building Regulations. Germany: University of Stuttgart; 2005, 2005. pp. 151-165

[3] ISO 834-1, Fire Resistance Tests - Elements of Building Construction – Part 1: General requirements, Chemin de Blandonnet 8, CP 401, 1214 Vernier, Geneva Switzerland; 1999

[4] EN 1991-1-2:2002, Eurocode 1: Actions on structures - Part 1-2: General actions - Actions on structures exposed to fire, CEN European Committee for Standardization, Management Centre: rue de Stassart, 36 B-1050 Brussels, Belgium, 2002

[5] EN 1992-1-2:2004, Eurocode 2, Design of concrete structures, Part 1–2: Structural fire design, CEN European Committee for Standardization, Management Centre: rue de Stassart, 36 B-1050 Brussels, Belgium, 2004

[6] Wade C. Method for Fire Engineering Design of Structural Concrete Beams and Floor Systems. Judgeford: BRANZ, The Resource Centre for Building Excellence; 1991

[7] American Concrete Institute. ACI / TMS 216.1–14, Standard Method for Determining Fire Resistance of Concrete and Masonry Construction Assemblies. Farmington Hills, Michigan, USA: American Concrete Institute; 2014

[8] ASTM International – ASTM. Standard Test Methods for Fire Tests of Building Construction and Materials (ASTM E119–14). Conshohocken, Pennsylvania, USA: ASTM International; 2014

[9] Lie TT, editor. Structural Fire Protection, (Technical Report No. 78). New York: American Society of Civil Engineers, USA; 1992

[10] Dzidic S, Kovacevic I, Kozlica S. Concrete Studies. Bosnia and Herzegovina: International BURCH University Sarajevo; 2018

[11] Dzidic S. Otpornost betonskih konstrukcija na pozar. Bosnia and Herzegovina: International BURCH University Sarajevo; 2015

[12] EN 1992-1-1:2004, Eurocode 2, Design of concrete structures, Part 1–1: General rules and rules for buildings, CEN European Committee for Standardization, Management Centre: rue de Stassart, 36 B-1050 Brussels, Belgium, 2004

Chapter 6

Low-Emission, Cementless Binders and Concrete: Future Proof Materials

Krystyna Rajczyk, Wiesław Kurdowski, Paweł Pichniarczyk and Grzegorz Janus

Abstract

In the chapter, the results of testing the possibility of obtainment of the high-performance binder and concrete based on alkali activated fly ash and other aluminosilicate wastes or industrial wastes are presented. The discussed topic is framed in terms of requirements and actions taken by the cement industry, which lead to a decrease in carbon dioxide emissions. In terms of carbon footprint, cementless, geopolymer concrete is determined as a more advanced material for civil engineering because of its ability to not contain the clinker component and still be able to reach, by its properties, the level of high-performance cementitious materials. In the presented paper, some of the properties of hardened geopolymer composite material are improved. It was established that the addition of thermally processed waste material, containing metakaolin, radically increases the strength and durability of geopolymer concrete. By means of completed research methods, such as X-ray diffraction (XRD), scanning electron microscopy (SEM) coupled with energy X-ray dispersion spectroscopy (EDS) and thermal analysis DTA-DTG, the influence on changes in the structure of hardened geopolymer material due to the increase in its strength and durability during low-temperature exposure is explained.

Keywords: alkaline activation, geopolymer binder, cementless concrete, low-emission materials, cementless binders

1. Introduction

Geopolymers are inorganic amorphous aluminosilicates. Geopolymers are made up of chains of AlO_4^{5-} and SiO_4^{4-} tetrahedra that are irregularly connected by these tetrahedra. These chains are arranged in layers, and between them, there are cations of sodium as well as groups of OH and H_2O. In this structure, there are periodic cations of aluminium, surrounded by four oxygen atoms [1].

The geopolymer binder binds and hardens as a result of many chemical reactions between aluminosilicate oxides in a strongly alkaline environment, forming three-dimensional polymer chains Si–O–Al–O [2, 3]. The geopolymerisation mechanism is highly dependent on the nature of alkaline activators as well as the chemical composition of precursors. The components used as precursors for geopolymerisation are low calcium aluminosilicates, e.g. silicious fly ash or blast-furnace slag. As shown in the figure, in addition to fly ash and granulated blast-furnace slag, raw materials such as metakaolin, halloysite and volcanic tuff are good precursors (**Figure 1**). According to Davidovits [2], the process of geopolymerisation consists of four main stages (dissolution, diffusion, polycondensation and hardening). This process was also described in a similar way by Glukhovsky [1]. De Silva et al. [3] suggested a three-step description of the geopolymerisation process (dissolution, agglomeration and polycondensation). Currently, one of the most important experts in the technology, microstructure and process of geopolymerisation of geopolymer materials is an Australian professor – John Provis, who promotes renaming geopolymers as Alkali-Activated Materials (AAM) [4].

Geopolymers are cementless, hardened, mechanically resistant materials with properties similar to natural stone or well-known concrete. The bonding process of such materials is different compared with the hydration process of cement, which involves the hydrolysis of calcium silicates and the formation of a hydrated C-S-H phase, with the simultaneous release of calcium hydroxide. The bonding process of geopolymers is slower compared with cement bonding. However, according to the literature, this time is sufficient to use the geopolymer binder to obtain concrete [2].

The preliminary research, held in the 1970s of the twentieth century in Poland, had led to development of geopolymer material by alkaline activation of granulated blast-furnace slag [5–8]. These studies were pioneering research in the country, which was a starting point for the numerous studies on geopolymer binder in other research centres such as Ł-ICIMB in Cracow, AGH in Cracow, Cracow University of Technology, and Poznan University of Technology. As it turned out, the volcanic tuff from Filipowice proved to be a good precursor for obtaining geopolymer materials [9].

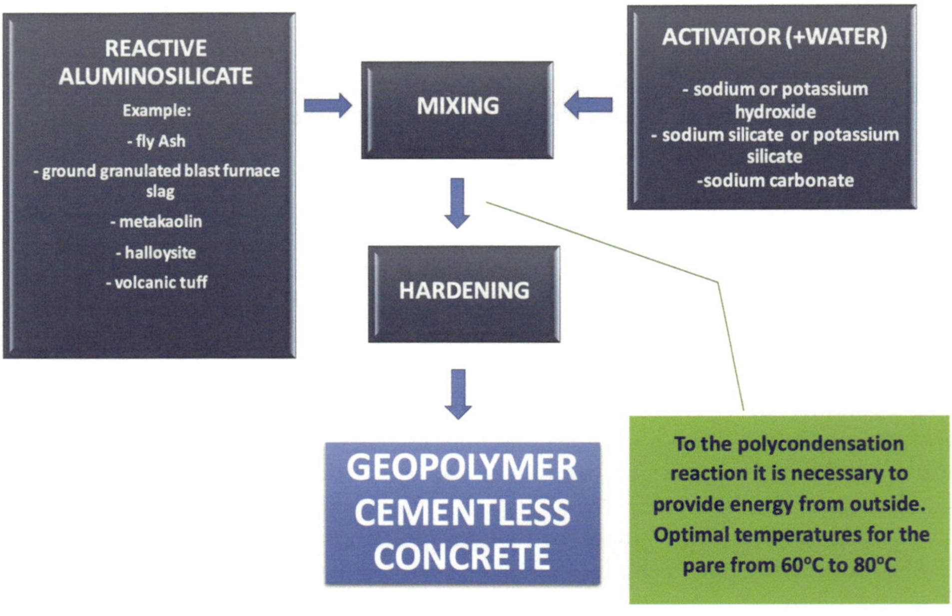

Figure 1.
Scheme for obtaining geopolymers.

In addition, the authors of the cited study have shown that post-mining waste such as coal shale [10], fly ash from lignite combustion [11] and calcined clay [12] can be an attractive material for the production of geopolymer binders.

The Ł-ICIMB (Łukasiewicz Research Network – Institute of Ceramics and Building Materials) has been conducting research on the use of fly ash from national power plants as a precursor for the production of geopolymer concrete [13–15]. Based on the XRD, SEM and thermal studies, it was found that the binding and hardening processes of cementless binder, based on fly ash, are the result of the dissolution of the active components of fly ash in a solution of sodium hydroxide. Obtained gel of silicoaluminates crystallises to form a hydrated sodium aluminosilicate of the sodalite or other type of zeolites such as: hydrosodalite, zeolite P, chabazite-Na and faujasite [16]. The resulting phases are durable and resistant to water and ambient environment, as confirmed by long-term strength tests [15, 17].

The degree of suitability of fly ash for geopolymerisation is determined by the SiO_2/Al_2O_3 ratio [18] and the CaO/SiO_2 ratio [19, 20] in fly ash, which determine the content of active components, which enter the solution under the influence of an alkaline activator [21]. According to the literature, most fly ashes have suitable properties, allowing the obtaining of binding cementless binders [18, 22]. However, as shown in [13, 15], the most advantageous fly ash allowing the production of high-strength binders and concrete is silicious micro-fly ash. The ability to convert the fly ash into geopolymer material is related mainly to the glassy phase composition, degree of gradation and loss of ignition [23–25].

Many studies have been conducted on the production of geopolymer materials based on fly ash [13, 17, 26–28]. The properties of hardened materials depend not only on the properties of the precursors, but also on the used activators. The most commonly used activators are sodium or potassium water glass, NaOH or KOH solution or a mixture of these compounds at experimentally determined proportions and concentrations [29–32]. The use of hydrothermal treatment (low-pressure steaming or autoclaving) accelerates the geopolymerisation reaction [16, 33, 34].

Geopolymer binder and concrete are widely considered to be low-carbon materials. Recently, many research centres around the world are undertaking research in the field of technology to obtain such materials. This has been included in the Roadmap of the European Cement Association (CEMBEREAU) as one of the options for reducing direct carbon dioxide emissions from the cement industry [35]. Taking into account that the cement industry is currently responsible for around 7% of anthropogenic CO_2 emissions [36], the proposal of the use of geopolymer binders in a partial exchange for high-energy and energy-consuming cement binders can significantly reduce the environmental burden and protect raw materials resources by increasing the use of waste materials and industrial waste. However, according to the literature review, most of the available studies have used either large quantities of alkaline activators or thermal processing in order to obtain geopolymer materials with properties comparable to or superior to traditional cement concrete. Such procedures raise production costs and energy consumption [37].

However, high cost of activators is a serious hindrance in the wide and industrial application of cementless geopolymer materials. This main reason hinders the competitiveness of alkali activated binders to Portland cement in spite of a much less energy-consuming of geopolymer binders. According to the data presented in [38], almost 60% of carbon dioxide emission from the process of producing geopolymer materials is associated with the production of alkaline activators. Therefore, some authors of publications on the production of geopolymer concrete believe that this technology does not lead to a significant reduction in carbon dioxide emission [39–41]. When making a reliable assessment of the environmental impact of the production and use of geopolymer materials, it is important to pay attention to the method of calculation and the data adopted by the authors. As Davidovits [42] explains, taking into account the life cycle of concrete, the production and use of geopolymer concrete lead to a reduction in CO_2 emissions. Furthermore, a majority of the authors believe that using geopolymer concrete instead of traditional cement concrete reduces CO_2 emission into the atmosphere [43–46].

The development of geopolymer technologies is justified because of the environmental and economic aspects, as well as the expectations of civil engineers. This technology is currently under research not only in the sphere of aggregate concrete but also in the technologies of obtaining lightweight geopolymer concrete [47, 48], self-compacting geopolymer concrete [43, 49] and even geopolymer foam concrete [44] or reinforced geopolymer concrete [27, 47].

The number of publications devoted to geopolymer concrete is constantly growing. In 2019–2020, it amounted to around 5200 and was almost twice as high as in 2017–2018 and more than eight times as high as in 2011–2012 [50]. However, only about 10% of the publications concern research on the aspects of construction and engineering applications [28], while the remainder is confined to laboratory-scale research [48, 51]. Despite many studies, the results of which can be observed in numerous publications [27, 28, 43, 44, 47–49, 51–54], geopolymer concrete has not received international appreciation as a construction material. It seems to be necessary to develop standards for the production of geopolymer concrete and procedures for designing geopolymer concrete structures [50].

An example of one of the first uses of geopolymer concrete is the building of the University of Queensland – Global Change Institute (GCI). The structure of this four-storey building is constructed from, among other, suspended slab floors and facade panels made of geopolymer concrete. The precursor used for the production of geopolymer concrete was a mixture of fly ash and blast-furnace slag [55]. In Australia, a widely developed technology is the production of sewer pipes, railway sleepers, burial crypts, culverts and wall panels from geopolymer concrete [56, 57]. Currently, one of the largest applications of geopolymer concrete is the element of the taxiway at Brisbane West Wellcamp Airport [58]. Long-term durability studies preceded the use of this concrete as a structural material [57]. However, according to Srividya et al. [50], in the current state of knowledge, it is necessary to further document the durability of such material under different exposure conditions and over a longer period of time. Regardless of ongoing research around the world, extensive literature reviews summarising scientific achievements in the field of geopolymer concretes have been published in recent years [37, 50].

The presented chapter presents the results of research on geopolymer materials obtained in the process of alkaline activation of fly ash and on geopolymer concrete with their participation. The optimal parameters of the activator quantity and the cure method for obtaining high-strength concrete were determined in previous works [13, 15]. Preliminary research carried out at the Ł-ICIMB on the development of an active additive for cement and concrete based on domestic kaolin resources has led to the establishment of optimal conditions for the thermal processing of raw clay for the conversion of the kaolin component into amorphous metakaolin [59]. The chapter also presents extensive studies and their results on increasing the durability of concrete in a low-temperature environment by enriching the concrete formula with kaolin mineral additive which were partially discussed in [15].

2. Research materials

Three types of fly ash were selected for the study: siliceous fly ash from hard coal combustion (FA1, FA2), fluidised fly ash from lignite combustion (FA3) and calcareous fly ash (FA4, FA5, FA6). FA2 ash, termed 'micro-fly ash' was selectively sampled. The FA5 and FA6 calcareous fly ashes were subjected to special treatment to improve some of their properties. The carbon fractions were extracted and rejected from the FA5 and FA6 ash. Moreover, the FA6 ash was additionally ground after the separation of the carbon fractions. **Figure 2** shows SEM image of the ashes used in the study.

Based on SEM image observations, it was found that the ashes used in the studies differed significantly in size and shape of the particles. While micro-fly ash (FA2) is characterised by fine oval-shaped particles (**Figure 2A**), calcareous fly ash is characterised by a more irregular shape (**Figure 2B**). The additional grinding of the calcareous ash after extraction of the carbon fractions (**Figure 2C**) resulted in the release of fine particles from conglomerates of larger ash particles as shown in **Figure 2D**. As a result of this procedure, very fine-grained ash was obtained. The results of the laser analysis obtained for FA2 and FA6 ashes are shown in the diagram (**Figure 3**).

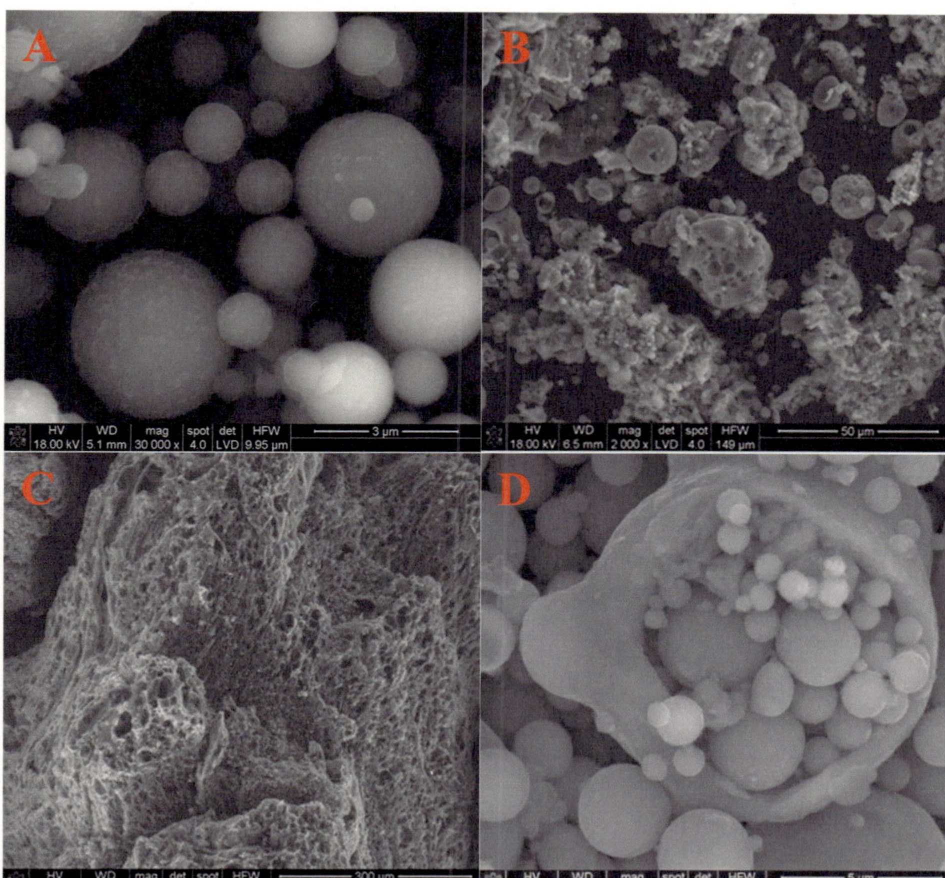

Figure 2.
SEM image of: A: 'micro-fly ash' - FA2, B: calcareous fly ash after the carbon extraction - FA5, C: extracted carbon particles from calcareous fly ash -FA4, and D: calcareous fly ash after additional grinding - FA6 [60].

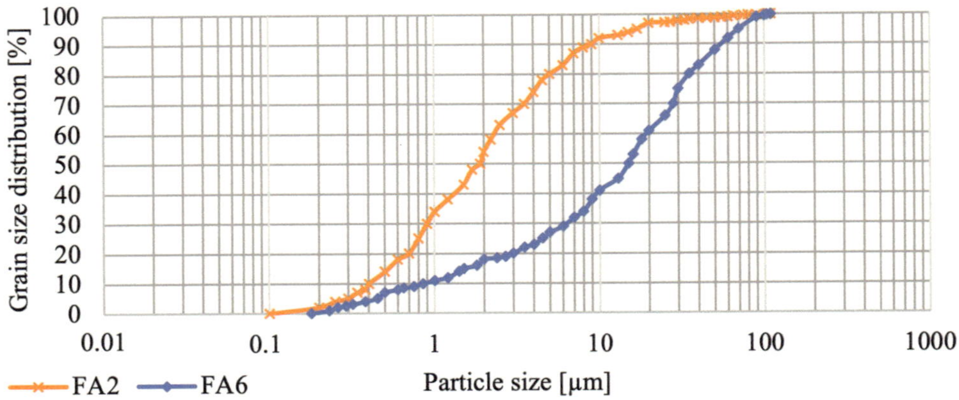

Figure 3.
Cumulative curve of particle size distribution of FA2 and FA6 fly ashes.

FA2 and FA6 ash particles showed significant differences in the granulometric distribution of these samples. FA2 fly ash is characterised by a predominant number of grains smaller than 10 μm (more than 90%). The maximum FA2 ash particle size was about 50 μm. The FA6 fly ash sample is characterised by larger particles. The percentage of particles smaller than 10 μm is about 40%, despite the additional grinding of the ash. The maximum particle size for this sample is about 100 μm. By analysing the particle size distribution curve of FA6 fly ash, it can be concluded that this material has a particle accumulation in the range of 10–60 μm and 0.5–1 μm.

As an additive to increase the durability of geopolymer concrete, a material containing a metakaolin component obtained by calcination of waste clay was used in the studies. The conditions of calcination were established on the basis of XRD studies and on the basis of the results of studies obtained using thermal analysis [13]. The phase composition of the clay used, as determined by XRD studies, is shown below.

The clay used contains significant amounts of calcite (**Figure 4**, **Table 1**). As can be seen from **Figure 4**, characteristic lines for calcite of considerable intensity are observed. The other components of the clay are kaolinite, quartz and a trace of minerals with a total content of less than 1%.

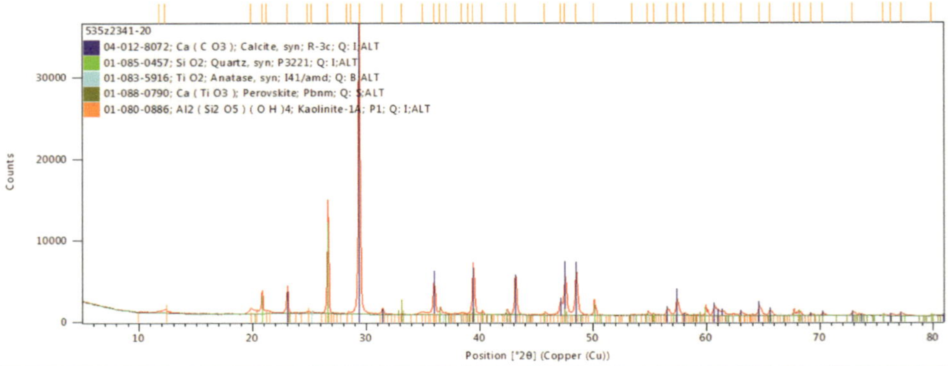

Figure 4.
XRD pattern of the raw waste clay.

Material	Quantitative contribution of phase composition [%]	Standard deviation [%]
Calcite	67.7	0.38
Kaolinite	17.9	0.42
Quartz	13.6	0.14
Anatase	0.40	0.06
Calcium titanate	0.40	0.10

Table 1.
Quantitative phase composition of the raw waste clay.

Based on DTA and DTG curves, dihydroxylation of kaolinite present in the waste clay begins in the temperature above 400°C, with the maximum of the endothermal effect at the temperature of 560°C (**Figure 5**). As a result of dihydroxylation, the metakaolin arises. According to Kurdowski [61], simultaneously with further temperature increase, the OH^- ions are eliminated and the structure of metakaolin is destroyed. The spinel phase of aluminosilicate and silicon is formed. Further temperature increase leads to conversion of these phases into mullite.

The DTA and DTG curves (**Figure 5**) also show an endothermic effect with a maximum temperature of 800°C associated with the decomposition of calcite. According to the accepted assumptions, the thermal activation temperature used in the waste clay studies should lead to the transformation of kaolinite into metakaolin, preventing its crystallisation into mullite. In addition, the thermal activation of clay assumes the lowest possible calcination temperature to prevent the complete decomposition of calcium carbonate in order to achieve the lowest possible level of CO_2 emissions. For the waste clay used in the study, the time of calcination was established to 2 hours and the optimum temperature of this process was determined as 700°C.

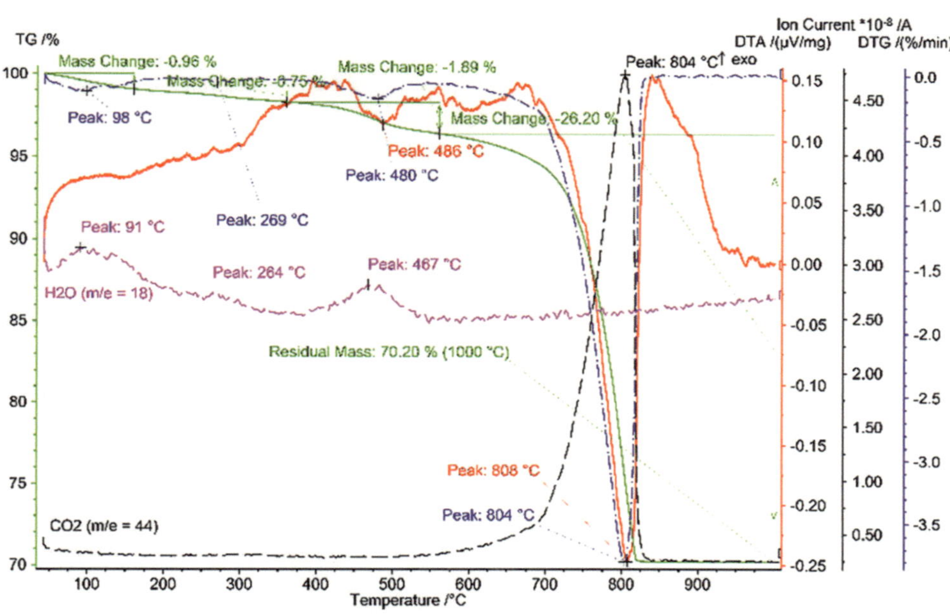

Figure 5.
DTA and DTG curves of the raw waste clay sample.

3. Geopolymer binder forms alkali-activated fly ash

The fly ash was activated using the 8 M NaOH solution. The ratio of the alkaline solution (as an activator) to the fly ash was determined experimentally. It was 0.5 or 0.6 depending on the applicable fly ash. The samples in the form of prisms with dimensions of 40 x 40 x 160 mm were prepared for testing. After ash paste was obtained, the samples were cured under low-pressure conditions at 80°C with an ongoing 24-hour heating-maturing-cooling cycle. The samples were then subjected to compressive strength tests, which are summarised in **Figure 6**.

In addition, to determine the effect of curing conditions of alkali-activated micro-fly ash (FA2), strength tests of hardening samples of geopolymer binder under ambient conditions were carried out. **Figure 7** shows the results of the compressive strength tests of the curing samples at room temperature and humidity. It has been observed that, without the use of low-pressure steam curing conditions, curing the geopolymer binder at room temperature (20**°C**) and air humidity of about 50% for a period of up to 56 days does not lead to the obtaining of high strength. Hardened geopolymer binder

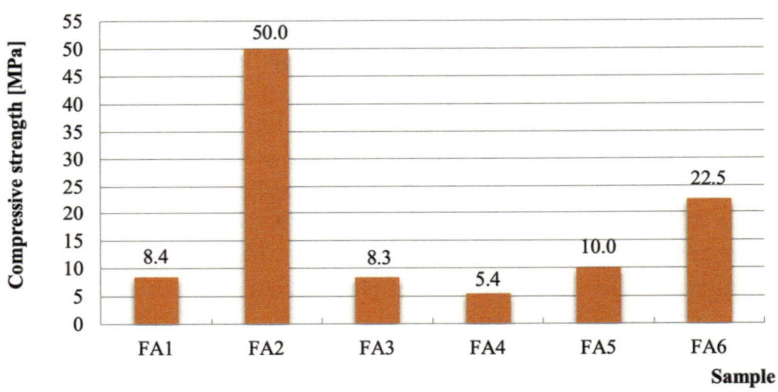

Figure 6.
Compressive strength of the fly ash pastes activated with the 8 moles/dm^3 NaOH solution and steam cured under low pressure at 80°C [13].

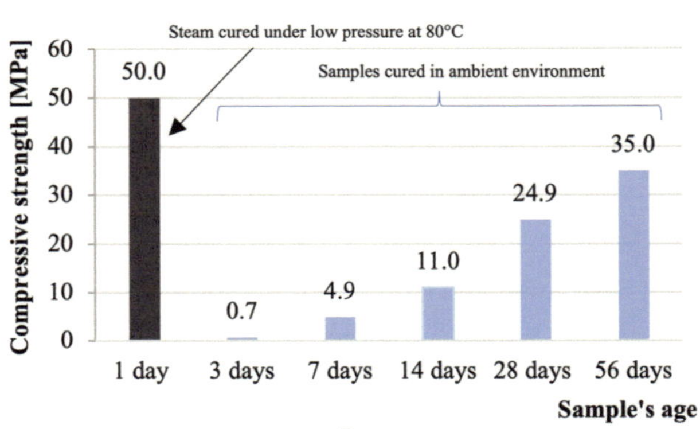

Figure 7.
The influence of the curing condition on the compressive strength of alkali activated fly ash FA2.

after this period obtained a compressive strength of 35.0 MPa, while immediately after the low-pressure steam process, it achieved a compressive strength of 50.0 MPa. In addition, white coating forms on samples that harden at room temperature, which does not occur when subjected to the low-pressure steam curing process.

Figure 8A shows the microstructure of the hardened geopolymer (FA2) characterised by the highest compressive strength. SEM image observations show a porous geopolymer structure formed between the fly ash particles under the influence of alkaline activation. The formed geopolymer phase adheres closely to the fly ash particles, forming a compact structure of hardened material (**Figure 8B**). A larger ash particle with a diameter of about 30 μm is visible in the test area.

The studies carried out at a higher magnification, together with the EDS analysis, made it possible to identify the composition of the formed compounds. At selected points of this sample, EDS analysis shows the formation of the geopolymer microstructure around the ash particles under the influence of alkaline activation with a variable Si/Al ratio (**Figure 8B** and **Table 2** p. 2, p. 3 and p. 4).

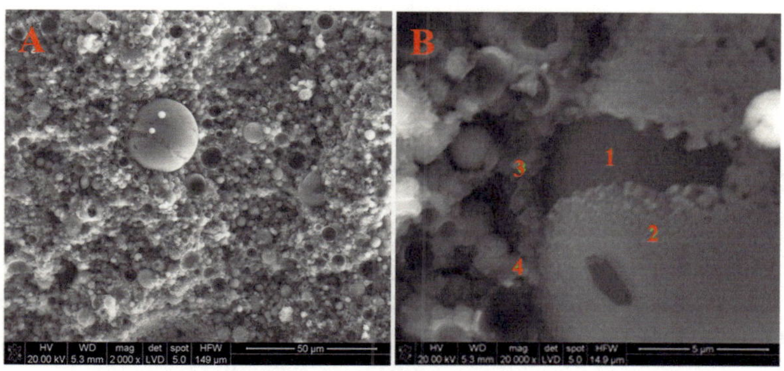

Figure 8.
SEM images of: A: of the hardened binder obtained from the FA2 (magnification 2000), and B: magnification 20,000 [13].

Point	Na	Mg	Al	Si	S	K	Ca	Ti	Fe	O
1	3.57	0.78	19.8	31.8	0.00	6.96	0.64	0.00	5.16	31.3
2	2.57	0.73	16,0	41.2	0.41	7.99	0.48	0.00	1.31	29.4
3	7.42	1.48	14.9	27.6	0.44	3.89	1.31	0.00	5.09	37.8
4	7.96	0.99	13.1	17.8	1.45	2.14	1.91	0.41	4.48	48.8

Table 2.
EDS analysis of selected points in the micro region (marked in **Figure 8B***) of the binder obtained from the FA2 ash (atomic percentage (%)) [13].*

According to XRD analysis, quartz and mullite derived from fly ash are present in the hardened geopolymer paste (**Figure 9**). Identified reflections for hydrosodalite, recorded on the diffractogram, prove the presence of this phase in the hardened matrix of the material obtained from FA2 fly ash in alkali activation process.

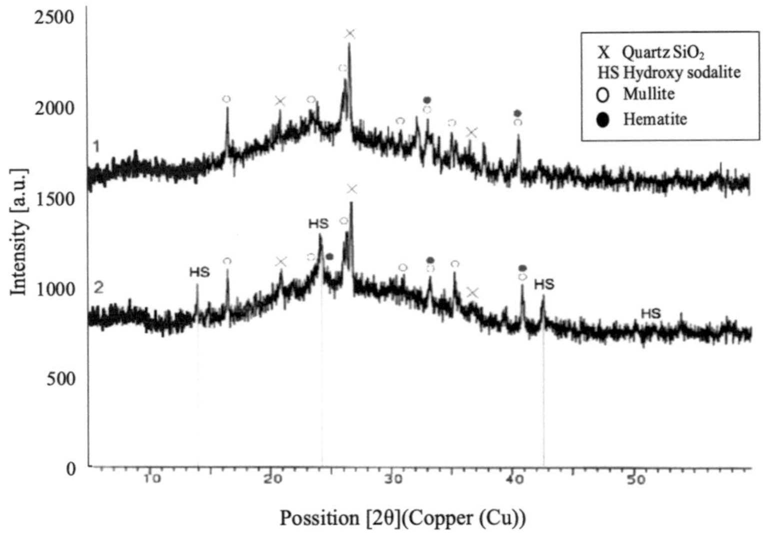

Figure 9.
XRD pattern: 1) – of the FA2 fly ash, 2) – of the hardened geopolymer paste obtained from the FA2 ash [15].

4. Development of geopolymer cementless concrete based on fly ash

The compositions of concrete mixes to obtain cementless geopolymer concrete were determined on the basis of the principles of designing ordinary concrete, assuming that in place of cement, the ashes selected for the study were used as the binder component in the concrete mix.

To prepare the reference concrete (B0FA1), the developed concrete mix was used, where the content of binder was 260 kg/m^3 in which 23% of cement was replaced by silicious fly ash (FA1). This fly ash is coherent with requirements from PN-EN 450 standard. Three aggregates' fractions were used: fine sand (0/2 mm), fine gravel 2/8 mm and gravel 8/16 mm. Apart from the micro fly ash (FA2), during the BFA2 + G mix preparation, the mineral additive – calcined at 700°C waste clay was added in place of 15% of micro fly ash FA2. The 8 mole/dm^3 NaOH solution was used as the activator, while for the BFA2 + G concrete mix, the mixture of 8 mole/dm^3 NaOH solution with water glass with an SiO_2/Na_2O molar modulus of 2.5 was used. **Table 3** shows the designations of the concrete samples, consisting of a member containing the type of initial components used to obtain the cementless geopolymer concrete.

5. Compressive strength test results of geopolymer concrete

The concrete mixes, after being cast into 100 x 100 x 100 mm cubes, were cured at 80°C under low-pressure steam conditions. In **Figure 10**, the compressive strength of tested geopolymer concretes hardened in 24 h low-pressure process is shown. From all tested concretes, which contain different fly ashes subjected to alkali activation by 8 mol NaOH solution, the highest compressive strength (26.6 MPa) was obtained by BFA2 concrete; however, it does not receive the value of compressive strength for reference concrete (30.8 MPa). BFA2 + G concrete with the addition of waste clay is characterised by exceptionally high compressive strength (73.5 MPa), exceeding the

Concrete designation	Binder	w/s	Kind of activator	Compressive strength [MPa]
B0 FA1	77% OPC + 23% FA1	0.5	—	30.8
BFA1	FA1	0.5	8 M NaOH	14.8
BFA2	FA2	0.5	8 M NaOH	26.6
BFA3	FA3	0.6	8 M NaOH	5.5
BFA6	FA6	0.6	8 M NaOH	10.6
BFA2 + G	85% FA2 + 15% calcined waste clay	0.6	8 M $NaOH/Na_2SiO_2$ = 1/2	73.5

Table 3.
List of materials used as a binder to obtain geopolymer concrete and kind of the activator [15].

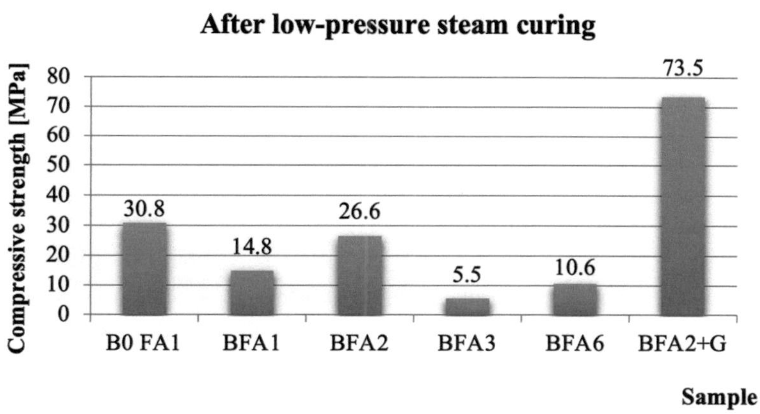

Figure 10.
Compressive strength of concrete samples after low-pressure steam curing [15].

strength of the cement reference concrete samples by almost 180%. In order to check the influence of the environmental conditions under which the samples were cured after steam curing, they were additionally stored in ambient conditions and in water at temperature of 20°C.

The concrete samples were additionally stored for a period of 28 days, both in ambient condition and in water, they did not lose the original strength, which they had obtained immediately after steam curing and obtained the high compressive strength, sometimes exceeding those of the initial sample. **Figure 11** shows the compressive strength of samples of the tested concretes after the low-pressure steam curing, stored additionally 28 days in water and in the ambient condition. To test the freeze-thaw resistance of the developed geopolymer cementless concretes, concrete samples were subjected to impact of low temperature. Unfortunately, not all samples of geopolymer concretes passed this test, which include of 150 freeze-thaw cycles. All samples prepared exclusively from fly ash as a binder, were completely destroyed, some of them after only 50 freeze-thaw cycles. The modification of the BFA2 concrete composition by replacing fly ash with 15% of a specially prepared additive—calcined waste clay—significantly improved the frost resistance of the geopolymer BFA2 + G concrete (**Figure 12**).

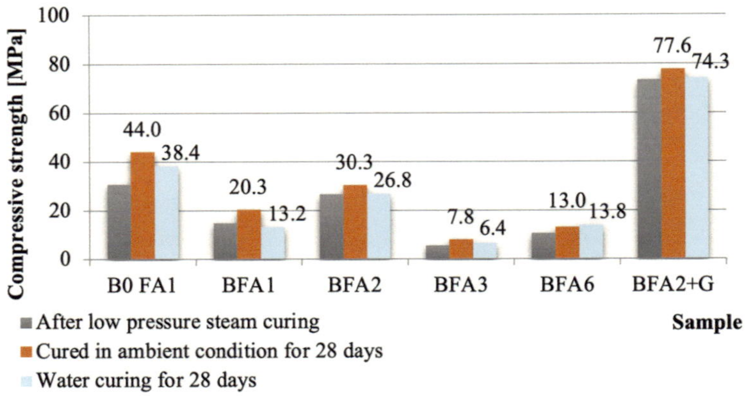

Figure 11.
Compressive strength of the concrete samples stored in ambient conditions and in water for 28 days after steam curing [15].

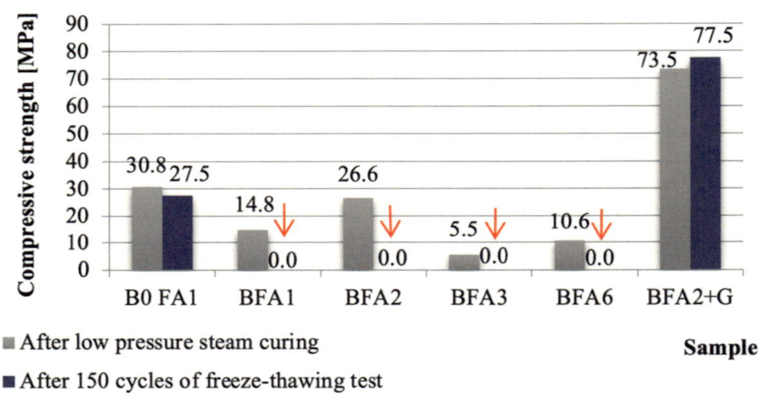

Figure 12.
Compressive strength of the geopolymer concretes after steam curing compared with the samples after frost resistance testing [15].

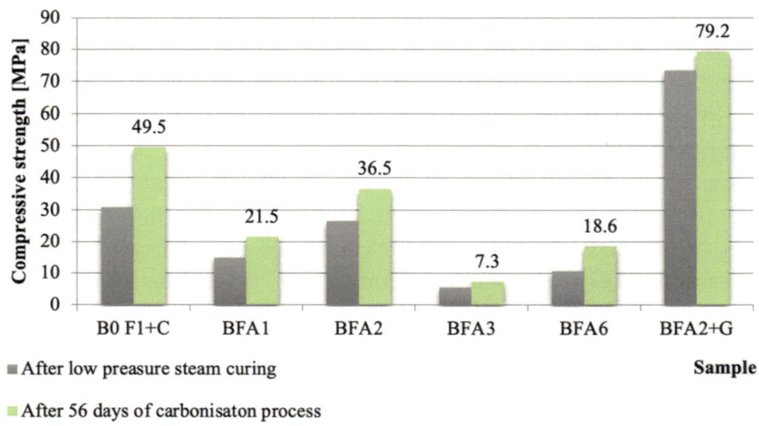

Figure 13.
Compressive strength of geopolymer concretes after low-pressure steam curing compared with the results of tests after carbonisation process.

Studies on the effect of carbonisation on the durability of the obtained geopolymer concrete were carried out in a test chamber. The pre-steamed samples were stored for another 28 days in ambient conditions and then were placed in a chamber, where they were subjected to the impact of high concentration of CO_2 (4%) for 56 days. The geopolymer concrete samples obtained higher strengths after the carbonisation process (**Figure 13**), which means that the samples after a period of preliminary steaming and further exposure show an increase in strength. The process of low-pressure curing does not lead to the achievement of the maximal strength of the material.

6. Microstructure of high-strength geopolymer concrete

The tests were carried out on the sample of the BFA2 + G concrete, which, due to the results achieved and the potential possibilities of use, as well as scientific value, raises universal interest. It is a cementless concrete made from secondary and waste materials which, after alkaline activation and subjected to low-pressure steam curing, was characterised by high compressive strength, high freeze-thaw resistance and durability both during curing in ambient conditions and in water and after conducting the carbonisation tests.

Figure 14 shows photograph exemplifying the structure of the fracture surface of the concrete sample after 3 years of storage in ambient conditions.

Further testing was carried out on the sample of the BFA2 + G, obtained by alkaline activation of fly ash with the addition of calcined clay. **Figure 15** shows the image under the scanning microscope of the microstructure of the fractured surface of the BFA2 + G concrete sample after 3 years of storage under ambient conditions. The test sample, extracted from the solidified matrix that bonds the aggregate particles, is not homogeneous. It contains partially reactivated fly ash particles. (**Figure 16A**) or more degraded fly ash particle (**Figure 16B**) which is a result of alkaline activation. The geopolymerisation products occur in a different morphology (**Figure 17** and **18**). The basic filler is a material formed by the reaction of the phase components of fly ash with an alkaline activator, which surrounds the unreacted fly ash particles (**Figure 15A**). The extender, formed by the alkaline activation of the components of the fly ash with the addition of calcined clay, is visible between the particles of the aggregate and the sand, which has a compact microstructure well bonded to the aggregate surface (**Figure 15A**). This substance consists mainly of silicon, aluminium and sodium. In the

Figure 14.
Macroscopic image of the structure of a geopolymer concrete sample.

Figure 15.
SEM images of selected areas of the BFA2 + G concrete sample: A: zeolite phase of alkaline fly ash activation products [15] and B: clusters of crystalline, hexagonal crystals [15].

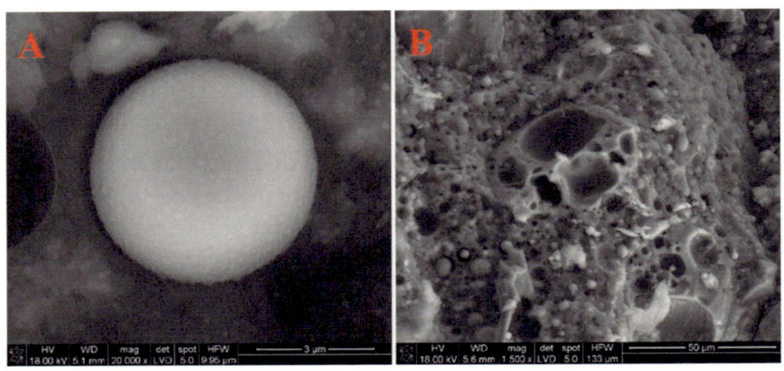

Figure 16.
SEM images of: A: partially reactivated fly ash particle with geopolymerisation products surrounding it and B: degraded fly ash particle as a result of alkaline activation.

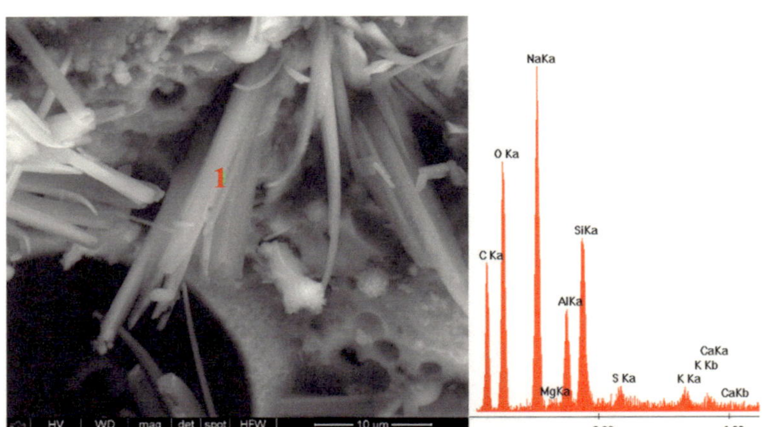

Figure 17.
The morphology of alkaline activation products in the form of needle-shaped, hexagonal crystals. They are the needle crystals of the zeolite group with high sodium content [15].

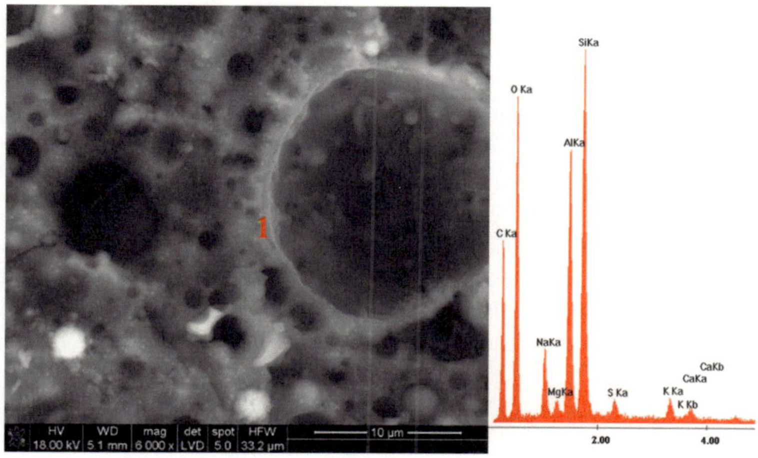

Figure 18.
SEM image of alkaline activation products formed around the surface of fly ash particles [15].

observed area of the sample, round, smooth ash particles are visible, which indicates that the alkaline reaction with the ash components occurs mainly in their surface layer (**Figure 18**). This is a normal phenomenon. They mainly contain silicon and aluminium and are characterised by a high sodium content. This component comes from the activator. In addition to the amorphous extender, clusters of crystallised hexagonal needle-like forms of the zeolite group are visible (**Figures 15B** and **17**).

Analysis of the phase composition conducted in the papier [15], as well as the SEM studies (**Figure 17** and **18**), indicates the content of crystalline phases in the test sample, originating from the base materials used. These include quartz, mullite, haematite and calcite. Based on the recorded in the papier [15] peaks with 'd' values of 4.7 Å and 3.64 Å on XRD pattern, it can be assumed that cancrinite ($Na_6Ca_2Al_6Si_6O_{24}(CO_3)_2$), a zeolite occurring in hexagonal needle-like forms, is present in the tested sample. The cluster of needle-like forms shown in **Figure 17**, with the EDS analysis carried out at point 1, indicates the possibility of the formation of this compound. The results of elemental X-ray microanalysis shown in the table show significant differences in the composition of geopolymer compounds formed as a result of the alkaline activation of FA2 ash with the addition of clay (**Table 4**).

The identified cancrinite (**Figure 17**, **Table 4**) contains significantly more sodium than the zeolite formed around the surface of the fly ash particle (**Figure 18**, **Table 4**). The content of this compound in concrete with other zeolites, such as hydrosodalite, formed by alkaline activation of fly ash, has been indicated by many researchers, as described by Davidovits [62] and Zhao et al. [63].

Probably, this phase crystallised from the amorphous precursors under the influence of CO_2 from the air, in which the developed geopolymer concrete was kept during the 3-year conditioning period. More likely, the crystallisation of this phase seems to be

Point	Na	Mg	Al	Si	S	K	Ca	Fe	O
1 on **Figure 17**	14.5	0.13	1.37	3.30	0.15	0.44	0.47	0.80	46.1
1 on **Figure 18**	3.00	0.59	8.02	12.2	0.64	1.04	0.66	1.21	33.5

Table 4.
EDS analyses of selected points in the micro region on **Figures 17** *and* **18** *(atomic percentage (%).*

caused by alkali activation of mineral components of calcined clay which includes apart from metakaolinite some amounts of undecomposed calcite. This thesis could be based according to tests carried out by Esaifan et al. [64], who proved the presence of both minerals: hydrosadalite and cancrinite by conducting the synthesis of those zeolites' minerals from calcite-containing clay under hydrothermal conditions. Calcite is a source of the necessary Ca^{2+} ions required to form cancrinite. This phase can crystallise already in a use of NaOH solution, not only by using a mixture of sodium silicate with NaOH solution [64]. In the results interpretation of XRD tests, the structure of hydrosadalite and cancrinite tested by Barnes et al. [65] turned out very helpful. According to these authors for those mineral phases on XRD pattern, many common diffraction peaks occur. The main identifying peaks for cancrinite unlike hydrosadalite are 4.67 Å and 3.24 Å.

7. Discussion

The presented work is based on the results of many years of studies carried out at Ł-ICIMB in Cracow, aiming the decrease of CO_2 emission from cement and concrete production. One of many ways of achieving this aim is the development of new, low-emission binders and concrete, which could complement the so-called classic materials such as cement and cement concrete and replace cement in certain areas of its application.

In the conducted research, focused on utilisation for obtaining the cementless binders and concrete, commonly available fly ashes from Polish power plants. According to conducted freeze-thaw tests, all samples of geopolymer concretes containing as a binder fly ash had been damaged after 150 freeze-thaw cycles. Even BFA2 concrete, despite high fresh compressive strength, failed this test. Especially low freeze-thaw resistance was demonstrated by geopolymer concretes based in alkali activated fluidised fly ash and untreated calcareous fly ash. After 50 cycles of freeze-thaw, those concretes were totally destroyed. The modification attempt of increasing the durability of concrete by addition of calcinated clay, as a partial replacement of fly ash, became a success. This test was conducted for the BFA2 concrete, which is characterised by the highest compressive strength among from tested concretes. Although it is known that the addition of metakaolin increases the strength of the fly ash based geopolymer [66], the results presented in this chapter are difficult to interpret unambiguously because, at the same time, the type of activator was changed, introducing 30% of water glass in place of sodium hydroxide. Consequently, the structure of the geopolymer filler of the hardened BFA2 + G geopolymer concrete is different compared with BFA2 concrete containing the same silicious micro fly ash from coal combustion.

The results obtained confirm the statements of numerous authors on the possibility of shaping the functional properties of geopolymer materials by selecting the suitable base materials, the type and quantity of activator, as well as hardening conditions. Currently, one of the factors determining the development of geopolymer in civil engineering will be the cost of material production. Therefore, it is advisable to look for cheaper, alternative activators. Reducing the carbon footprint is also an important element. It is worth paying attention to the comparison of the CO_2 emission reduction and the calculation of the carbon footprint in relation to the cement material. Only the analysis carried out for products with identical technical characteristics can be considered a conclusive comparison.

Taking into account the tendency to reduce coal burning in the energy sector, it should be noted the possibility of the limitation the extraction of the best for geopolymer concrete, silicious fly ash. This will be followed by work on the identification of other aluminosilicate precursors from both waste materials and industrial waste. A suitable example is an attempt to obtain appropriately transformed calcareous fly ash as the main component of cementless geopolymer concretes. The presented results of the use of refined calcareous fly ash, through the separation of carbon fractions and additional grinding, confirmed the possibility of obtaining cementless geopolymer concrete from such a material. A similar effect was obtained by Blaszczyński and Król [67], who, as one of the few authors, also obtained a geopolymer binder from tentatively refined calcareous ashes by removing carbon particles and pre-grinding them. Ash of this kind has not found application as an ingredient for cement production, so there are great opportunities for its acquisition.

For the widespread production of structural materials based on geopolymer binders and concrete, it is necessary to carry out long-term durability studies and develop guidelines for testing procedures and standard requirements so that the geopolymer material "native to ancient Egypt" with the new name AAM (Alkali-Activated Material) becomes the material of the future [4].

8. Conclusion

Part of the research results on the development of geopolymer concrete based on waste aluminosilicate materials, in particular fly ash, were published in the journal Cement Wapno Beton [15]. Results of tests, showed in this chapter, make a significant contribution, which led to interpret the phenomenon that occurs during alkali activation of chosen aluminosilicates industrial wastes. It has extended current knowledge in a field of geopolymer materials.

Among the fly ashes described, selected for research as base materials for obtaining cementless binders and geopolymer concretes, the best properties were demonstrated by fine-grained, specially selected siliceous fly ash from hard coal combustion—so-called micro fly ash.

It was demonstrated that a large role in the formation of the microstructure of the hardened geopolymer concrete showing high durability under the tested conditions of exposure to both water and ambient condition and low temperatures can be attributed to the modification of developed geopolymer microstructure by the presence of metakaolin and calcium carbonate in the calcined at 700°C clay [14, 60, 64, 66]. This modification resulted by the occurrence in the geopolymer microstructure, in which needle-like forms were identified using a scanning microscope, which, following other authors, can be attributed to the formation of cancrinite [64, 68].

Acknowledgements

We would like to thank the Cement Wapno Beton foundation for disseminating the results of our research in the article entitled 'Microstructure and properties of geopolymers formed in the alkali activation process of fly ash', which resulted in our invitation to take part in the preparation of the chapter of the book entitled 'Advanced Cement-Based Materials'. This research was accomplished within statutory activity at Research Network – Institute of Ceramics and Building Materials by the team represented Materials Engineering Research Group.

Conflict of interest

The authors declare no conflict of interest.

Author details

Krystyna Rajczyk[1*], Wiesław Kurdowski[2], Paweł Pichniarczyk[1] and Grzegorz Janus[1]

1 Łukasiewicz Research Network – Institute of Ceramics and Building Materials, Kraków, Poland

2 Cement Wapno Beton Foundation, Kraków, Poland

*Address all correspondence to: krystyna.rajczyk@icimb.lukasiewicz.gov.pl

References

[1] Glukhovsky VD, Rostovskaya GS, Ramana G. High strength slag-alkaline cements. In: 7th ICCC. Vol. 3. Paris; 1980. pp. 164-168

[2] Davidovits J. Geopolymer: Inorganic polymeric new materials. Journal of Thermal Analysis and Calorimetry. 1991;**37**:1633-1656. DOI: 10.1007/bf01912193

[3] De Silva P, Sagoe-Crenstil K, Sirivivatnanon V. Kinetics of geopolymerization: Role of Al_2O_3 and SiO_2. Cement and Concrete Research. 2007;**37**(4):512-518. DOI: 10.1016/j.cemconres.2007.01.003

[4] Górski M. Geopolimers – Ecologic future materials native to ancient Egypt (in polish). In: Inżynier Budownictwa. 2021

[5] Brylicki W. Modification of alkali activated slag pastes by use of mineral admixtures. In: 10th ICCC Göteborg. 1997. pp. 4-7

[6] Małolepszy J. Hydration and properties of alkali activated slag binder (in polish). In: Zeszyty Naukowe AGH - Ceramika. 1989

[7] Deja J. Phase composition of slag pastes activated by alkali (in polish). Cement Wapno Beton. 2005;**3**:127-137

[8] Derdacka-Grzymek A, Małolepszy J, Brylicki W, Deja J. Hydraulic binder (in polish). Patent PL nr 271470

[9] Mikuła J, Łach M. Production and properties of geopolymers based on volcanic tuff (in polish). Inżynieria Materiałowa. 2014;**35**(3):270-276

[10] Stefańska A, Łach M, Mikuła J. Geopolimers as an example of the possibility of waste utilisation (in polish), Nowoczesne technologie XXI w. – przegląd, trendy i badania. Volume 1. Wydawnictwo Naukowe TYGIEL sp. z o. o. 2019

[11] Mikuła J, Łach M. Geopolimers – New, environmental friendly, alternative for. Concretebased on Portland cement. Examples of implementation and test results. In: Pro-Ecological Solutions in a Range of Production (in Polish). Kraków: Politechnika Krakowska; 2014

[12] Łach M, Gado RA, Marczyk J, Ziejewska C, Doğan-Sağlamtimur N, Mikuła J, et al. Process design for a production of sustainable materials from post-production clay. Materials. 2021;**14**(4):953. DOI: 10.3390/ma14040953

[13] Rajczyk K, Giergiczny E, Szota M, Microstructure and Properties of Hardened Geopolimer Binders from Fly Ash (in Polish). Prace Instytutu Ceramiki i Materiałów Budowlanych 2015

[14] Rajczyk K, Janus G, Kaliciak A, Brukhanska D. Cementless concrete from kaolinite and alumino-silicate waste materials. In: Book of Abstracts of the 13th Conference for Young Scientist in Ceramics. Serbia: Novi Sad; 2019. pp. 147-148

[15] Rajczyk K, Janus G. Microstructure and properties of geopolymers formed in the alkali activation process of fly ash. Cement Wapno Beton. 2021;**26**(4):279-293. DOI: 10.32047/CWB.2021.26.4.2

[16] Garcia-Lodeiro J, Maltceva O, Palomo A, Fernandes-Jiménez A. Hybrid alkaline cements: Part I. fundamentals. Romanian. Journal of Materials. 2012;**42**:330-335

[17] Derdacka-Grzymek A, Stok A. Cementless binder from fly ash (in polish). Cement Wapno Gips. 1980;**8-9**:220-222

[18] Fernández-Jiménez A, Palomo A. Characterisation of fly ashes. Potential reactivity as alkaline cements. Fuel. 2003;**82**(18):2259-2265. DOI: 10.1016/S0016-2361(03)00194-7

[19] Yip CK, Lukey GC, Van Deventer JSJ. The coexistence of geopolymeric gel and calcium silicate hydrate at the early stage of alkaline activation. Cement and Concrete Research. 2005;**35**(9):1688-1697. DOI: 10.1016/j.cemconres.2004.10.042

[20] Antoni A, Satria J, Sugiarto A, Hardjito D. Effect of variability of fly ash obtained from the same source on the characteristics of geopolymer. MATEC Web of Conferences. 2017;**97**:01026. DOI: 10.1051/matecconf/20179701026

[21] Wattimena OK, Antoni A, Hardjito D. A review on the effect of fly ash characteristics and their variations on the synthesis of fly ash based geopolymer. AIP Conference Proceedings. 2017;**1887**(1):020041. DOI: 10.1063/1.5003524

[22] Van Jaarsveld JGS, Van Deventer JSJ, Lorenzen L. The potential use of geopolymeric materials to immobilize toxic metals: Part I. Theory and applications. Minerals Engineering. 1997;**10**(7):659-669. DOI: 10.1016/S0892-6875(97)00046-0

[23] Temuujin J, Van Riessen A, Williams R. Influence of calcium compounds on the mechanical properties of fly ash geopolymer pastes. Journal of Hazardous Materials. 2009;**167**(1-3):82-88

[24] Antoni A, Wibiatma WS, Hardjito D. Factors affecting the setting time of fly ash – Based geopolymer. Materials Science Forum. 2016;**841**:90-97. DOI: 10.4028/www.scientific.net/MSF.841.90

[25] Škvarla J, Sisol M, Botula J, Kolesárová M, Krinická I. The potential use of fly ash with a high content of unburned carbon in geopolymers. Acta Geodynamica et Geomaterialia. 2011;**8**(2):123-132

[26] Andini S, Cioffi R, Colangelo F, Grieco T, Montagnaro F, Santoro L. Coal fly ash as raw material for the manufacture of geopolymer-based products. Waste Management. 2008;**28**(2):416-423. DOI: 10.1016/j.wasman.2007.02.001

[27] Wallah SE, Rangan BV. Low-Calcium Fly Ash-Based Geopolymer Concrete: Long-Term Properties. Australia: University of Technology Perth; 2006

[28] Radhakrishnan S, Jeyalakshmi R, Selvan KG, Rajamane N. Fly ash/slag geopolymer technology development and deployment in construction and infrastructure industry: India's perspective. IJESE. 2017;**10**:757-765

[29] Gougazeh M. Geopolymers from Jordanian metakaolin: Influence of chemical and mineralogical compositions on the development of mechanical properties. Jordan Journal. 2019;7(2):236-257

[30] Risdanareni PUPUT, Ekaputri JJ, Al Bakri Abdullah MM. Effect of alkaline activator ratio to mechanical properties of geopolymer concrete with trass as filler. Applied Mechanics and Materials. 2015;**754**:406-412. DOI: 10.4028/www.scientific.net/AMM.754-755.406

[31] Parthiban K, Saravanarajamohan K, Shobana S, Bhaskar AA. Effect of replacement of slag on the mechanical properties of fly ash based geopolymer

concrete. International Journal of Engineering & Technology. 2013;**5**(3):2555-2559

[32] Lloyd N, Rangan V. Geopolymer concrete with fly ash. In: Proceedings of the Second International Conference on Sustainable Construction Materials and Technologies. 2010. pp. 1493-1504

[33] Palomo A, Grutzeck M. W, Blanco MT. Alkali activated fly ashes: A cement for the future. Cement and Concrete Research. 1999;**29**:1323- 1329. DOI: 10.1016/S0008-8846(98)00243-9

[34] Gultekin A, Ramyar K. Effect of curing type on microstructure and compressive strength of geopolymer mortars. Ceramics International. 2022;**48**(11):16156-16172. DOI: 10.1016/j.ceramint.2022.02.163

[35] Novel Cements. Available from: https://lowcarboneconomy.cembureau.eu/5-parallel-routes/resource-efficiency/novel-cements/ [Accessed: April 14, 2022]

[36] Maddalena R, Roberts JJ, Hamilton A. Can Portland cement be replaced by low-carbon alternative materials? A study on the thermal properties and carbon emissions of innovative cements. Journal of Cleaner Production. 2018;**186**:933-942. DOI: 10.1016/j.jclepro.2018.02.138

[37] Alhawat M, Ashour A, Yildirim G, Aldemir A, Sahmaran M. Properties of geopolymers sourced from construction and demolition waste: A review. Journal of Building Engineering. 2022;**50**:104104. DOI: 10.1016/j.jobe.2022.104104

[38] Turner LK, Collins FG. Carbon dioxide equivalent (CO_2^{-e}) emissions: A comparison between geopolymer and OPC cement concrete. Construction and Building Materials. 2013;**43**:125-130. DOI: 10.1016/j.conbuildmat.2013.01.023

[39] Habert G, de Espinose E, Lacaillerie JB, Roussel N. An environmental evaluation of geopolymer based concrete production: Reviewing current research trends. Journal of Cleaner Production. 2011;**19**:1229-1238

[40] Nehdi ML, Yassine A. Mitigating Portland cement CO_2 emissions using alkali-activated materials: System dynamics model. Materials. 2020;**13**(20):4685. DOI: 10.3390/ma13204685

[41] Grzeszczyk S. The truth about the geopolymers. Cement Wapno Beton. 2021;**26**(2):101-108

[42] Davidovits J. False values on CO_2 emission for geopolymer cement/concrete published in scientific papers. Technical Paper, 24, Geopolymer Institute Library. 2015

[43] Demie S, Nuruddin MF, Shafiq N. Effects of micro-structure characteristics of interfacial transition zone on the compressive strength of self-compacting geopolymer concrete. Construction and Building Materials. 2013;**41**:91-98. DOI: 10.1016/j.conbuildmat.2012.11.067

[44] Singh NB. Foamed geopolymer concrete. Materials today. Proceedings. 2018;**5**:15243-15252. DOI: 10.1016/j.matpr.2018.05.002

[45] Farhan NA, Sheikh MN, Hadi MNS. Experimental investigation on the effect of corrosion on the bond between reinforcing steel bars and fibre reinforced geopolymer concrete. Structure. 2018;**14**:251-261. DOI: 10.1016/j.istruc.2018.03.013

[46] Zhang Z, Provis JL, Reid A, Wang H. Geopolymer foam concrete: An emerging

material for sustainable construction. Construction and Building Materials. 2014;**56**:113-127. DOI: 10.1016/j.conbuildmat.2014.01.081

[47] Wongsa A, Sata V, Nuaklong P. Chindaprasirt P, use of crushed clay brick and pumice aggregates in lightweight geopolymer concrete. Construction and Building Materials. 2018;**188**:1025-1034. DOI: 10.1016/j.conbuildmat.2018.08.176

[48] Islam A, Alengaram UJ, Jumaat MZ, Ghazali NB, Yusoff S, Bashar II. Influence of steel fibers on the mechanical properties and impact resistance of lightweight geopolymer concrete. Construction and Building Materials. 2017;**152**:964-977. DOI: 10.1016/j.conbuildmat.2017.06.092

[49] Muttashara HL, Ariffinb MAM, Husseinc MN, Hussinb MW, Ishaqb SB. Self-compacting geopolymer concrete with spend garnet as sand replacement. Journal of Building Engineering. 2018;**15**:85-94. DOI: 10.1016/j.jobe.2017.10.007

[50] Srividya T, Kannan Rajkumar PR, Sivasakthi M, Sujitha A, Jeyalakshmi R. A state-of-the-art on development of geopolymer concrete and its field applications. Case Studies in Construction Materials. 2022;**16**:e00812. DOI: 10.1016/j.cscm.2021.e00812

[51] Ganesan N, Abraham R, Raj SD. Durability characteristics of steel fibre reinforced geopolymer concrete. Construction and Building Materials. 2015;**93**:471-476. DOI: 10.1016/j.conbuildmat.2015.06.014

[52] Amin M, Elsakhawy Y, Abu El-hassan K, Abdelsalam B. Behavior evaluation of sustainable high strength geopolymer concrete based on fly ash, metakaolin, and slag, case studies. Construction Materials. 2022;**16**:e00976. DOI: 10.1016/j.cscm.2022.e00976

[53] Al-Majidi MH, Lampropoulos A, Cundy AB. Steel fibre reinforced geopolymer concrete (SFRGC) with improved microstructure and enhanced fibre-matrix interfacial properties. Construction and Building Materials. 2017;**139**:286-307. DOI: 10.1016/j.conbuildmat.2017.02.045

[54] İpek S. Macro and micro characteristics of eco-friendly fly ash-based geopolymer composites made of different types of recycled sand. Journal of Building Engineering. 2022;**52**:104431. DOI: 10.1016/j.jobe.2022.104431

[55] Bligh R, Glasby T. Development of Geopolymer Precast Floor Panels for the Global Change Institute at University of Queensland. Available from: http://www.asec 2014.org.nz/Presentations/PDFs/Paper%20037%20Development%20of%20Geopolymer%20Precast%20Floor%20panels%20for%20the%20Global%20Change%20Institute%20at%20University%20of%20Queensland.pdf. [Accessed: April 5, 2022]

[56] Gourley JT. Geopolymers in Australia. Journal of the Australian Ceramic Society. 2014;**50**(1):102-110

[57] Glasby T, Day J, Genrich R, Kemp M. Commercial scale geopolymer concrete construction. In: The Saudi International Building and Constructions Technology Conference 2015, Riyadh. 2015

[58] Glasby T, Day J, Genrich R, Aldred J. EFC Geopolymer Concrete Aircraft Pavements at Brisbane West Wellcamp Airport, Concrete 2015 Conference. Melbourne; 2015

[59] Rajczyk K, Giergiczny E. Metakaolin as an active pozzolanic additive (in polish). In: Presiding of the 2nd

Scientific Technical Conference of Kaolinite Raw Materials on the Threshold of the XXI Century. Leśna; 1998

[60] Rajczyk K. Fly ash from fluidized bed combustion and the possibilities of their refinement (in polish). WYDIS 2012 Opole

[61] Kurdowski W. Chemical Basis of Mineral Building Materials and Their Properties (in polish). Kraków: Stowarzyszenie Producentów Cementu; 2018

[62] Davidovits J. Geopolymer Chemistry and Applications. 4th ed. France: Institut Géopolymère; 2015

[63] Zhao Y, Zhao X, Xie H, Jiang Y. Synthese of cancrinite from alkalifused ash by "dry method". Advanced Materials Research. 2011;**236-238**:676-679. DOI: 10.4028/www.scientific.net/AMR.236-238.676

[64] Esaifan M, Warr LN, Grathoff G, Meyer T, Schafmeister MT, Kruth A, et al. Synthesis of hydroxy-sodalite/cancrinite zeolites from calcite-bearing kaolin for the removal of heavy metal ions in aqueous media. Minerals. 2019;**9**(484):1-13. DOI: 10.3390/min9080484

[65] Barnes MC, Addai-Mensah J, Gerson AR. A methodology for quantifying sodalite and cancrinite phase mixtures and the kinetics of the sodalite to cancrinite phase transformation. Microporous and Mesoporous Materials. 1999;**31**(3):303-319. DOI: 10.1016/S1387-1811(99)00080-3

[66] Barbhuiya S, Pang E. Strength and microstructure of geopolymer based on fly ash and metakaolin. Materials. 2022;**15**:3732. DOI: 10.3390/ma15103732

[67] Błaszczyński TZ, Król MR. Properties of aluminosilicates based on calcareous fly ash (in polish). Przegląd budowlany. 2019;**7-8**:46-55

[68] Passos F, Castro D, Ferreira K, Simões K, Bertolino L, Barbato C, et al. Synthesis and characterization of sodalite and cancrinite from kaolin. In: Characterization of Minerals, Metals, and Materials. Cham: Springer International Publishing; 2017. pp. 279-288. DOI: 10.1007/978-3-319-51382-9_31

Chapter 7

Physicochemical Behavior of Concretes Admixed with Water-Based Polymers (PMC: Polymer-Modified Concrete)

José Atílio Fritz Fidel Rocco, Rene Francisco Boschi Gonçalves and Marcela Galizia Domingues

Abstract

Portland cement concrete remains the most important of the materials used due to its mechanical properties combined with its low cost and ease of obtaining. However, conventional concrete structure suffers aging with gradual increase in porosity. This implies limitations and leads to the development of modifications to improve its general characteristics, like the so-called concrete modified by polymers that resulted in improvements in mechanical and microstructural properties, in relation to the original concrete. Concrete (cement)-polymer composites are the materials, which are made by replacing a part or all the cement hydrate binders of conventional mortar or concrete with polymers and by strengthening the cement hydrate binder with polymers. The composites are generally classified into three types by the principles of their process technology: 1. Polymer-modified mortar and concrete; 2. Polymer mortar and concrete; and 3. Polymer impregnated mortar and concrete. In this study, we evaluated the effect of the addition of the acrylic additive in the hydration reaction, comparing the result between the cement paste samples with and without added polymer. Beyond mechanical tests, such as compression resistance and traction, various thermal analysis techniques were used. The results showed great modifications on hydration process and mechanical behavior of PMC.

Keywords: physicochemical behavior, water-based polymers, concrete admixture, PMC-polymer modified cement, hydration reaction

1. Introduction

In recent years, technical innovations in the construction industry have progressed considerably, and the research and development of high-performance and multifunctional construction materials have been actively pursued to cope with the innovations. Because of this worldwide interest in concrete (cement), polymer has become stronger.

Concrete (cement)-polymer composites are the materials that are made by replacing a part or all of the cement hydrate binder of conventional mortar or concrete with polymers and by strengthening the cement hydrate binder with polymers.

The concrete (cement)-polymer composites are generally classified into the following three types by the principles of their process technology:

1. Polymer-modified (or cement) mortar (PMM or PCM) and concrete (PMC or PCC)

2. Polymer mortar (PM) and concrete (PC)

3. Polymer-impregnated mortar (PIM) and concrete (PIC).

In this study, we considered only type 1—polymer-modified (or cement) mortar.

The Portland cement was modified by an acrylic polymer latex addition. The properties of the mortars and cements that were obtained by adding acrylic latex to Portland cement were an improvement in the mechanical properties and some changes in the hydration process. This addition alters the early-age hydration kinetics of polymer-modified cement. In other words, the physicochemical behavior of *concretes admixed with water-based polymers (polymer-modified concrete)*.

Our experiments showed that the acrylic polymer retards the hydration process of the Portland cement. These changes were observed in different ways: mechanical properties and thermal analysis (TG/DTG—thermogravimetry and differential thermogravimetry; DSC—differential scanning calorimetry and semi-adiabatic calorimetry).

In the construction industry, many technical innovations have emerged in the latter years that have benefited the development of multifunctional and high-performance construction. Can be cited as examples the works buried as tunnels, buildings intelligent, and even lunar bases projects. In this way it is important, more and more, the development of new building materials that enable the care of these new high-performance requirements and include the environmentally friendly feature.

Portland cement concrete remains the most important of the materials used due to its mechanical properties combined with its low cost and ease of obtaining, which determine their advantages in relation, for example, to steel or other materials. However, conventional concrete has its limitations. Recent research shows that their structure suffers with aging, with a gradual increase in porosity as a whole [1]. This implies limitations and leads to the development of modifications to improve its general characteristics. One of these modifications led to the so-called concrete modified by polymers that, among other changes, resulted in improvements in mechanical and microstructural properties compared to the unlimited original concrete.

The concrete and mortar-modified mortars are increasingly used in applications such as tunnel coatings, reservoirs, roads, floors, repairs in old coatings, gluing for ceramic tiles, waterproofing, and chemical barriers.

In addition to meeting these new high-performance requirements already mentioned, the Portland cement concrete is exposed to a series of chemical species that can attack its microstructure. These chemical species can degrade concrete by dissolving its soluble constituents, such as calcium hydroxide. The dissolution of hydroxide calcium, in turn, leads to an increase in porosity by increasing the phenomena of percolation and diffusivity in this same microstructure.

Portland cement concrete can be modified by the addition of latices (aqueous polymeric dispersions), also known as latexes, changing their rheological and mechanical properties [2, 3]. Since 1950 in the USA, acrylic latex, defined as an anionic aqueous dispersion of an acrylic copolymer, has been studied as a modifier of mortar and Portland cement concrete with the objective of changing its original mechanical properties [3]. This modification with acrylics, which in general can be defined as a family of resins resulting from the acrylic acid polymerization, results in changes in the properties of healed concrete.

The rheological properties (fluidity), mechanical (compression resistance, traction, and abrasion), and permeability, which can be measured through resistance to ion penetration chloride [3]. In fresh concrete [3], for example, the addition of latex causes plasticizing rheological changes, besides reducing the WC factor (water/cement), which is a determining factor in the quality of the concrete.

The mechanism by which polymers interact with the main elements of Portland cement, such as silicates and aluminides, during the hydration reaction is of great importance in the study of its resistance to degradation. Various analysis techniques can be employed for the study of these interactions in the modified Portland cement microstructure. In addition to the traditional mechanical tests, the use of, among others, the scanning electron microscope [4], X-ray diffraction [4, 5], scattering of X-rays at low angles [6], spreading neutrons [7, 8], thermal analyzers [5, 9], and others can be mentioned.

In this work, we employ, beyond mechanical essays as resistance to compression and traction, the thermal analysis techniques, TG and DTG. In the present study, an effect evaluation of the addition of acrylic additive in the hydration reaction [10], comparing the result between the cement paste samples with and without added polymer. Beyond mechanical tests, such as compression resistance and traction, various thermal analysis techniques were used.

2. The Portland cement

Portland cement, as it is commonly known, is a mixture of compounds produced from the thermal decomposition of the limestone ($CaCO3$) and from clay at high temperatures ranging from 1400 to 1600°C with the addition of calcium sulfate after cooling. The production of Portland cement begins with the extraction of the limestone, which is then broken into small pieces by gigantic grinders. The ground limestone is then mixed with the clay, sand, and iron ore, apparently forming a homogeneous powder. However, this powder is microscopically heterogeneous [11, 12]. This resulting mixture is subjected to heat treatment at the temperatures mentioned above to obtain the clinker. By Portland cement clinker, the material is understood to be synthesized and pelletized, resulting from the calcination to approximately 1450°C of a proper limestone and clay mixture and eventually corrective components of silica, alumina, and even iron nature, employed to ensure the composition chemistry of the mixture within specific limits [11, 12]. This clinker, after cooling, is mixed with calcium sulfate for the correction of the catch time [11, 12], resulting in Portland cement itself.

3. Portland cement hydration: kinetic and mechanistic aspects

Portland cement contact with water triggers a series of reactive processes that lead to hydrated products resulting in a dense and stable microstructure. An important particularity of these complex reactions is that the initial reagents are in powder form.

The compounds present in Portland cement are anhydrous, but when put in contact with water, they react with it, forming hydrated products. Moisturizing cement Portland consists of the transformation of more soluble anhydrous compounds into compounds less soluble hydrated. During hydration, there is the formation of a gel layer in the grain of the original compounds, so that, in the transition zone (a zone intermediate between the primary crystal and the gel), the solution is supersaturated in relation to the hydrated compounds. The variations in concentration of solute and water generate a gradient of concentration, originating an osmotic pressure that will bring about the rupture of the gel, exposing new areas of the anhydrous compound to water action [12].

Two are the classic theories that seek to explain the hydration of Portland cement: Le Chatelier and Michaelis [12].

In Le Chatelier, crystallization hardening is explained by the shelter of crystals that are formed by the crystallization of a supersaturated solution of compounds that are hydrated but less soluble than the anhydrous. The binder, in definable terms, is a system of unstable anhydrous constituents that, in the presence of water, tend to give a system of stable hydrated constituents.

According to the colloidal theory of Michaelis, the hydration of the Portland cement gives rise to a supersaturated solution with the formation of crystals in needles and hexagonal reeds. There is formation of a hydrated monocalcic silicate, which gives rise to a gel colloidal of gellected mass, which imprints the crystals; the gel continues to drink water, the dough hardens and waterproof.

In this work, the percolation theory was used, where hydration is divided into three distinct phases: induction, nucleation, growth, and diffusion. More details of this theory will be presented below.

The study of hydration is important with the purpose of understanding, throughout its extension, as reactions occur throughout the process called cure. Studies of the hydration kinetics of each of the constituents of cement are found in the literature in isolation and usually serve as a reference [13–15]. Generally, the reactions that better represent the hydration process are mentioned below in a notation of the cement industry [16]:

1. $2C_3S + 6H \rightarrow C_3S_2H_3 + 3CH$
2. $2C_2S + 4H \rightarrow C_3S_2H_3 + CH$
3. $2C_3A + 21H \rightarrow C_4AH_{13} + C_2AH_8$
4. $C_4AH_{13} + C_2AH_8 \rightarrow 2C_3AH_6 + 9H$
5. $C_3A + 3CSH_2 + 26H \rightarrow C_6AS_3H_{32}$
6. $2C_3A + C_6AS_3H_{32} + 4H \rightarrow 3C_4ASH_{12}$
7. $C_4AF + 3CSH_2 + 29H \rightarrow C_6(A,F)S_3H_{32} + (A,F)H_3$
8. $C_4AF + C_6(A,F)S_3H_{32} + 7H \rightarrow 3C_4(A,F)SH_{12} + (A,F)H_3$

It is interesting to note that in terms of reactivity with water, these different phases of minerals that make up Portland cement behave differently. The work performed by

Jolicoeur and Simard [16] shows that the relative reactivity of the different mineral phases with water is usually given as C3A > C3S > C2S > C4AF (the absolute reactivities vary considerably depending on the degree of metal ion substitution in the phases and their crystal structure). Accordingly, the aluminate phases and their hydration products play an important role in the early hydration processes. In a general, simplifying sense, the early (O-l h) behavior of hydrating cements is governed by reactions of the aluminate phases, particularly C3A; the setting and early strength development behavior is mostly dependent on the hydration of silicates, particularly C3S. Because of the high reactivity of calcium aluminate and the undesirable properties of some of the products formed (e.g., hexagonal C-AH), the aluminate hydration reaction is carried out in the presence of sulfate ions. The latter provide control of the reaction rate through the formation of mixed aluminate sulfate products, namely ettringite (AFt) and monosulfoaluminate (AFm), i.e., reactions (5) and (6), respectively. Calcium sulfate added to the clinker can thus be viewed as a first 'chemical admixture' used to control the nature and properties of the aluminate hydration products. Sulfates thus play a crucial role in cement hydration, and the influence of chemical admixtures on any process involving sulfates may be expected to be significant.

The overall process of cement hydration and setting results from a combination of solution processes, interfacial phenomena, and solid-state reactions. To help visualize the influence of admixtures on cement hydration, it is useful to recall the main events of the hydration process and the time scale over which they occur. A schematic representation of the evolution of a Portland cement hydration reaction with time is reproduced in **Figure 1** (adapted from Jolicoeur and Simard-16). The latter identifies five distinct stages, the boundaries of which are determined by sharp variations in

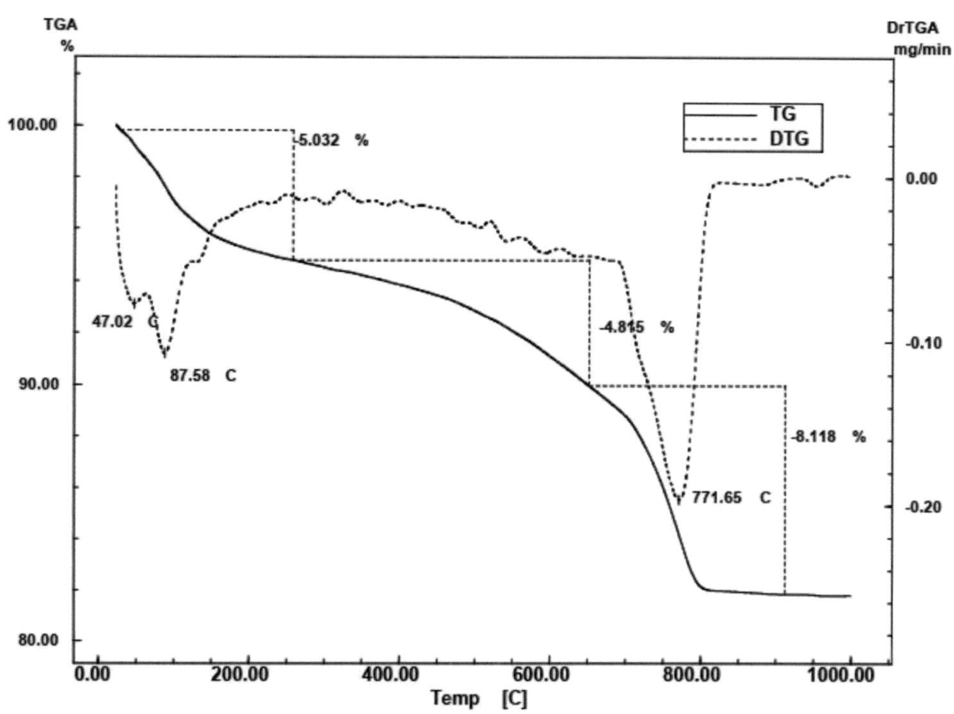

Figure 1.
TG/DTG curves obtained with a heating rate of 10 K/min and a dynamic air atmosphere of 50 mm/min for an unmodified Portland cement sample with a hydration age of 28 days and a water/cement factor of 0.45.

a reaction parameter, typically the heat flux measured as a function of time. These five stages correspond, respectively, to (times shown are approximate): I. Initial hydration processes (O–15 min); II. Induction period or lag phase (15 min–4 h); III. Acceleration and setting (4–8 h); IV. Deceleration and hardening (8–24 h); V. Curing (O–28 days).

Tricalcic silicate (C3S) is more reactive than dicalcic silicate (C2S), which, in turn, is more reactive than tricalcic aluminate (C3A), which has, approximately, the same reactivity as tetracalcic ironuminate (C4AF). According to these authors, complex reactions involving the mentioned species of chemicals cause the formation of a microstructure containing hydrated calcium silicates, represented by C-S-H; hydroxide in calcium, represented by CH; plaster; monosulfoalumines; and endothelium (a mixture of alumings and sulfates).

From the point of view of the reaction stoichiometry, according to the literature [16], for the Portland Type I (ASTM) cement, a water/cement ratio, known as the "water/cement factor (W/C), of 0.3 by weight is sufficient for the complete hydration of all phases of minerals that make up this same cement. In practice, however, this is not what occurs because the rheological characteristics such as viscosity should also be considered, which involve the processing of cementitious materials.

4. Portland cement samples evaluation

Figures 1 and **2** show the TG/DTG curves obtained for the cement samples unmodified and modified by the addition of latex, respectively. For the unmodified sample, two main stages of mass loss were observed. The first one can be attributed to dehydration that occurs from room temperature to approximately 600°C, with a total mass variation of 9.8%. At this stage, it was observed that between 25 and 200°C the mass loss is 5.5% and the DTG curve showed two distinct peaks at 45 and 87°C, indicating that in this material the water molecules are linked in different ways.

A second mass loss, which occurs between 600 and 800° C, corresponds to the release of 8.1% of CO_2 due to the thermal decomposition of 18.4% $CaCO_3$ present in the starting material. **Figure 2** shows the presence of the polymer, whose pyrolysis peak occurs at 350.19°C. The temperature peak at 91.85°C indicates that after adding the polymer, the water combines differently because the peak temperature without the addition of polymer is 87.58°C.

The Portland cement hydration process is exothermic [17] and has a great influence on the final properties of mortars and concrete obtained. The temperature variations that occur inside large masses of concrete are the subject of study by many researchers.

One of the ways of studying the kinetics of this hydration and verifying the influence of factors such as the polymer concentration in relation to the weight of cement [18], among others, is by monitoring the heat generated over time by semi-adiabatic calorimetry. In this technique, the heat of reaction released due to cement hydration, modified or not, is recorded in the form of temperature under semi-adiabatic conditions. This technique allows the monitoring of temperature variations that occur in the hydration of Portland cement by the addition of latex. As can be seen, for example, in **Figure 3**.

The temperature versus time curves were obtained from 0 to 24 hours and plotted in graphs where the changes that occurred in the phase known as nucleation and growth of the hydration reaction were compared.

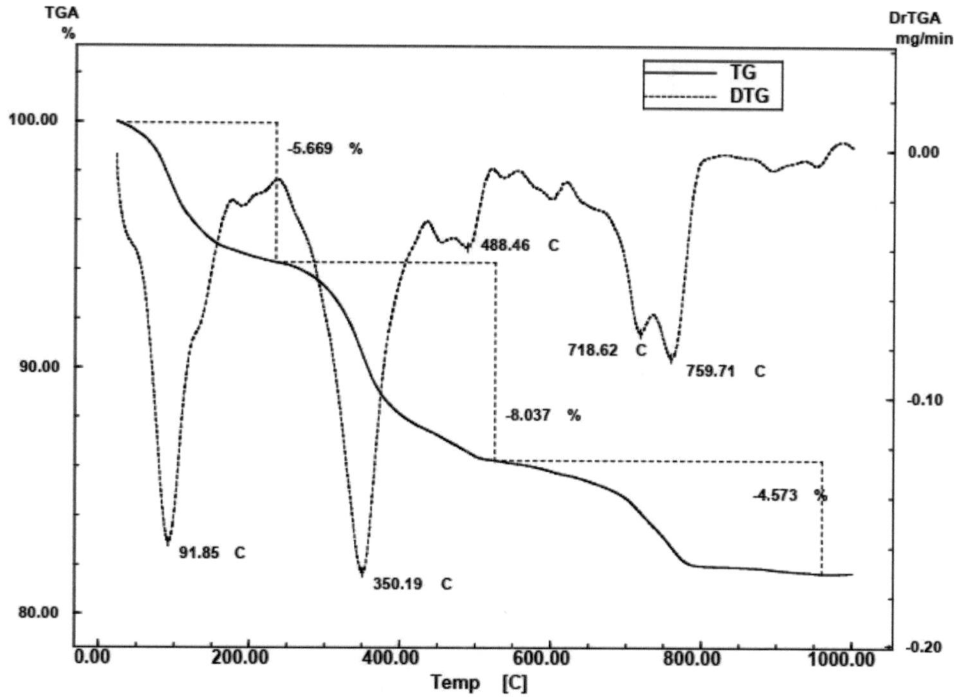

Figure 2.
TG/DTG curves obtained with a heating rate of 10 K/min and a dynamic air atmosphere of 50 mm/min of a modified Portland cement sample with the addition of 20% acrylic acid, a hydration age of 28 days, and a water/cement factor of 0.45.

Due to the characteristics of the semi-adiabatic calorimeter used in the test, the nucleation and growth phases were isolated because they present the greatest manifestation of heat in relation to the dormant and diffusion phases in the hydration reaction. The linear regressions of each one of the hydration temperature curves of the Portland cement paste as a function of time show how the addition of the acrylic additive, in terms of concentration, alters the kinetics of the hydration reaction.

Figure 3 is important because, in a qualitative way, the following hydration phases of Portland cement can be observed:

a. An induction period that starts at time t = 0 hours and ends at time t = 2 hours, under the conditions of the experiment. Also known as the dormant period, this is the phase in which the cement constituents undergo irreversible dissolution in water.

b. Nucleation and growth that begin at the end of the induction period and last until time t = 11 hours. It is in this phase that most of the hydrates and calcium hydroxide are formed, and most of the heat of the hydration process is released.

c. Diffusion, which starts right after the end of the nucleation and growth phase, represented in the graph by the peak temperature of 70°C and is limited by the speed with which water can diffuse through the crystalline microstructure formed to find and react with the remaining silicates and aluminates.

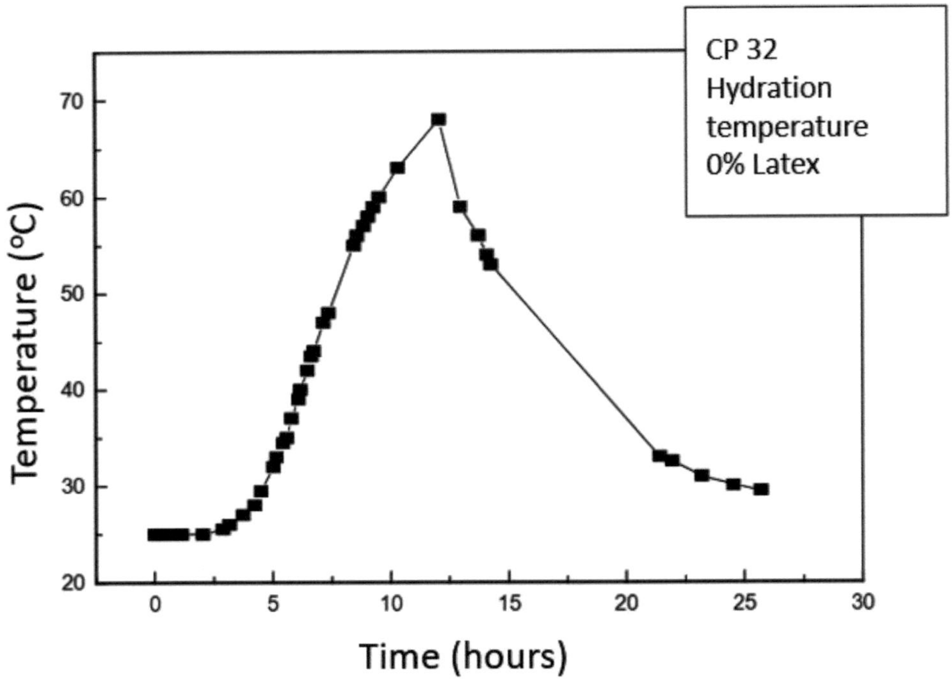

Figure 3.
Hydration temperature as a function of time for unmodified Portland cement paste (0% latex addition).

The semi-adiabatic calorimetry, unlike the adiabatic one, allows the clear observation of the phase change between nucleation and growth and of diffusion. This is because, as there is a relative loss of heat from the system to the environment in the phase change due to the drop in the rate of heat release, there is a sharp drop in temperature as observed in **Figure 3** (a peak of 70°C).

Once the hydration temperature of the unmodified Portland cement paste had been monitored under semi-adiabatic conditions, the same experiment was carried out for samples of Portland cement pastes modified by the addition of different levels of acrylic latex, as shown in **Figure 4.**

From the observation of the graph in **Figure 4**, it can be seen how the addition of the lattice, in terms of a percentage increase, changes the generally accepted reaction mechanism for the phases or periods:

a. Induction. There is a small change in the final time of this period depending on the increase in the percentage of incorporated latex content.

b. Nucleation and growth. It is the most affected phase and is undergoing major changes, mainly in terms of heat generated;

c. Diffusion. Directly affected due to the greater difficulty for water to come into contact with the unreacted chemical species of cement after the nucleation and growth phases.

In general, unmodified cement obtains the highest peak temperature and therefore generates the most heat, which decays as the polymer concentration increases. This is

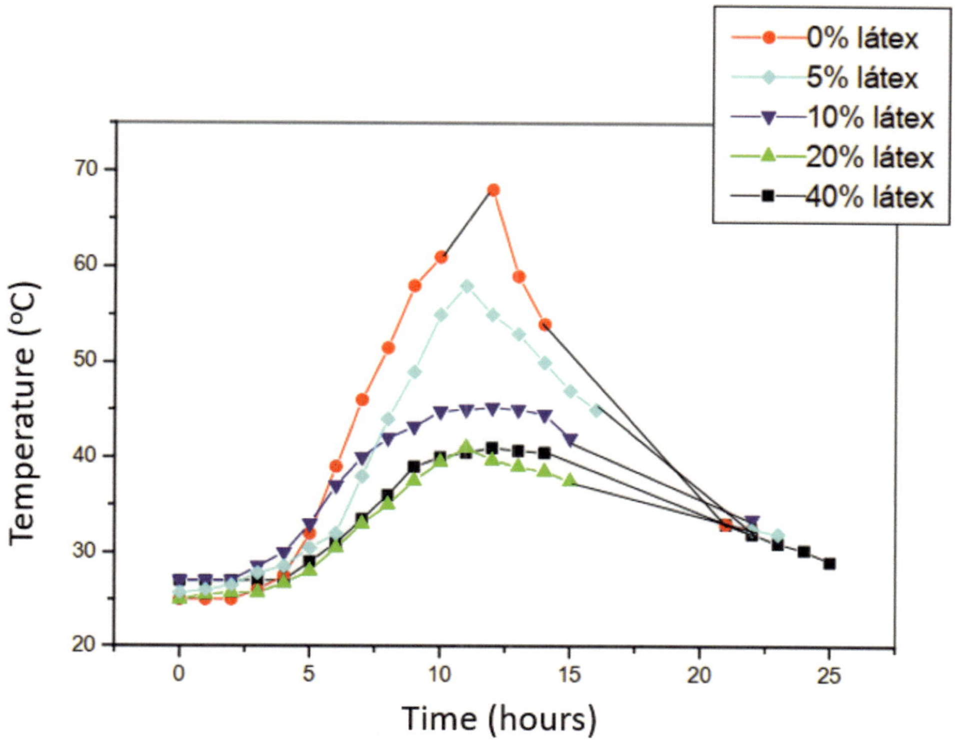

Figure 4.
Hydration temperature versus time as a function of modifications by the addition of acrylic latex to Portland cement.

explained because the acrylic latex acts as a retarder in the Portland cement hydration process.

On the other hand, its use in the modification of Portland cement leads to a decrease in the heat generated during hydration. This can be especially beneficial for situations where the addition, due to the complexity of the part or construction system, has a limiting factor in the quality of the concrete obtained in the release of heat.

The small increase in the induction period, the period in which the mortar and/or concrete are transported and cast, is also beneficial from a quality point of view, as it allows a longer time between manufacture and final molding.

Possible interactions between the polymeric latex and the cement particles occur, leading to this delay in the hydration process. Among these interactions, the following are presumed: reduction of the initial dissolution capacity of the chemical compounds of cement in water due to the presence of latex; formation of a polymeric film around the grains of the chemical compounds of cement, preventing the reaction with water; formation of complex compounds between polymeric particles and calcium ions in solution, with consequent reduction in the formation of calcium hydroxide crystals.

5. Conclusions

The modification of Portland cement by the addition of acrylic latex alters the behavior of its physical–chemical properties. When considering the hydration kinetics, the results obtained through the use of thermal analysis techniques allow us to conclude that the addition of polymers modifies the hydration kinetics of Portland cement at the considered initial ages. Probably due to the encapsulation effect caused by the involvement of Portland cement powder unreacted by the polymer, the contact and subsequent reaction with water becomes difficult.

Also, due to a decrease in porosity as the empty spaces tend to be filled by the polymeric chain, there is an increase in the difficulty of water diffusion in the resulting microstructure during the diffusion phase.

It can be noted by observing the TG and DTG curves that water tends to have two distinct behaviors: Without the addition of polymer, a peak is observed at 87°C; with the addition of 20% latex, the water peak shifts to 92.35°C.

This behavior indicates that water, when the polymer is added, tends to bind in the microstructure of hydrated Portland cement in a different way. The same behavior is verified for $CaCO_3$, whose peak for the unmodified Portland cement paste is observed at 771.65°C. When modified by the addition of latex, this peak shifts to a temperature of 760.34°C. Another peak at 716.68°C may indicate the presence of water bound to the C-S-H microstructure.

Author details

José Atílio Fritz Fidel Rocco*, Rene Francisco Boschi Gonçalves and Marcela Galizia Domingues
Departamento de Química, Instituto Tecnológico de Aeronáutica, São José dos Campos, SP, Brazil

*Address all correspondence to: friz@ita.br

References

[1] Odler I, Chen Y. Investigation on the aging of hydrated Tricalcium silicate and Portland cement pastes. Cement and Concrete Research. 1995;**25**(5):919-923

[2] Ohama Y. Recent Progress in concrete-polymer composites. Advancement Based Materials. 1997;**5**:31-40

[3] Lavelle JA, Wright PE. Acrylic modifiers for cement. Rohm and Hass Research Laboratories. 1974;**XXIV**(2):18

[4] Bentz DP, Stutzman PE. SEM analysis and computer modelling of hydration of Portland cement particles. In: DeHayes SM, Stark D, editors. Petrography of Cementitious Materials, ASTM STP 1215. Philadelphia: American Society for Testing Materials; 1995

[5] Ollitrault-Fichet R, Gauthier C, Clamen G, Boch P. Microstructural aspects in a polymer-modified cement. Cement and Concrete Research. 1998;**28**(12):1687-1693

[6] Winslow DN, Diamond S. Specific surface of hardened Portland cement paste as determined by small-angle X-ray scattering. Journal of American Ceramic Society. 1974;**57**(5):193

[7] Fitzgerald SA, Neumann DA, Rush JJ, Bentz DP, Livingstone RA. In situ quasi-elastic neutron scattering study of the hydration of Tricalcium silicate. Chemistry of Materials. 1998;**10**(1):397-402

[8] Berliner R, Popovici M, Herwig K, Jennings HM, Thomas J. Neutron scattering studies of hydrating cement pastes. Physica B. Elsevier. 1998;**241 – 243**:1237-1239

[9] Sha W, O'Neill EA, Guo Z. Differential scanning calorimetry study ofordinary Portland cement. Cement and Concrete Research. 1999;**29**:1487-1489

[10] Zeng S, Short NR, Page CL. Early – Age hydration kinetics of polymer-modified cement. Advances in Cement Research. 1996;**8**(29):1-9

[11] Aquino RA. Caracterização e Estudo Termoanalítico de Clínquer Produzido em Indústrias Brasileiras, Tese de Doutorado. Instituto de Química, Universidade de São Paulo; 1996. Available from: https://www.ipen.br/biblioteca/teses/21964.pdf

[12] Petrucci EGR. Concreto de Cimento Portland. Porto Alegre: Editora Globo; 1979

[13] Taylor JC, Aldridge LP. Full-profile Rietveld quantitative XRD analysis of Portland cement: Standard XRD profiles for the major phase tricalcium silicate (C3S:3CaOSiO2). Powder Diffraction Journal. 1993;**8**(3):138-144

[14] Douy A, Gervais M. Crystallization of amorphous in the Calcia-alumina system: A differential scanning calorimetry XRD study. Journal of the American Ceramic Society. 2000;**83**(1):70-76

[15] Bentz DP, Garboczi EJ, Haecker CJ, Jensen OM. Effects of cement particle size distribution on performance properties of Portland cement-based materials. Cement and Concrete Research. 1999;**29**(10):1663-1671

[16] Jolicoer C, Simard M. Chemical admixture-cement interactions: Phenomenology and Physico-chemical concepts. Cement and Concrete Composites. 1998;**28**:87-101

[17] Bentz DP, Waller V, De Larrard F. Prediction of adiabatic temperature rise

In conventional and high-performance concrete using a 3-D microstructure model. Cement and Concrete Research. 1998;**28**(2):285-297

[18] Ohama Y. Polymer-based admixtures. Cement and Concrete Composites. 1998;**20**:189-212

Section 2

Design and Optimization of Reinforced Concrete

Chapter 8

Borderless Optimization Method Genetic Algorithm

Mustafa Kaya

Abstract

Genetic algorithms, a stochastic research method, are Darwin's "survival of the best." Development of biological systems based on the principle of "survival of the fittest" emerged with the adaptation of the process to the computer environment. In genetic algorithms, processes executed are stored in computer memory similar to natural populations performed on units. Today, there are many linear and nonlinear solutions for optimization problems. The method has been developed. Designing optimization problems with these method variables is assumed to be continuous. But this acceptance does not always valid. In some problems, there are too many design variables and constraints for traditional optimization in solving such problems. The use of this method sometimes gives erroneous results or the solution time is too long. Since genetic algorithms are an intuitive method, a given may not find the optimum result for the problem. However, it optimizes the problems that cannot be solved by known methods or the solution time of the problem is too long. Initial nonlinear optimization genetic algorithms applied to the problems of traveling salesman, workshop scheduling, and syllabus preparation, such as complex optimization problems, have been successfully implemented.

Keywords: evolutionary algorithms, artificial intelligence, genetic algorithms, stohastic research method, binary coding

1. Introduction

Genetic algorithms, a stochastic research method, emerged by adapting the development process of biological systems to the computer environment. Operations using genetic algorithms are performed on units stored in computer memory, similar to natural populations. Today, many linear or nonlinear methods have been developed for the solution of optimization problems. In these problems, it is accepted that the design variables are continuous. However, this acceptance is not always valid. Due to a large number of design variables and constraints in some problems, the use of traditional optimization methods in solving such problems sometimes gives erroneous results or the solution time is too long. Genetic algorithms may not find the optimum result for a given problem. However, it gives values very close to the optimum values for problems that cannot be solved by known methods or whose solution time increases exponentially with the solution of the problem. Genetic algorithms have been successfully applied to complex optimization problems, such as traveling salesman, workshop planning, and curriculum preparation. The basic principles of genetic algorithms were first introduced by Holland [1]. The discovery of the crossover operator by Holland played a major role in the development of genetic algorithms. In his studies, Holland was driven by the idea of developing species that can adapt very well to their environment with the effect of sexual intercourse, reproduction, and natural selection. The first work in the literature on genetic algorithms is John Holland's Machine Learning. The study by David E. Goldberg, a student of Holland, who was later influenced by this study, on the control of gas pipelines proved that genetic algorithms can have practical uses. The problems for which genetic algorithms are most suitable are those that cannot be solved by traditional methods or whose solution time increases exponentially with the size of the problem. Until today, it has been tried to solve problems in different fields using genetic algorithms. Some of these fields of study are optimization, automatic programming, machine learning, economics, immune systems, population genetics, evolution, and learning and social systems.

2. Structure and basic principles of genetic algorithms

In the first step of the genetic algorithms, an initial collection of subsets of all possible solutions is obtained. Each member of the community is coded as an individual. Each individual is biologically equivalent to a chromosome. Every individual in the community has a fitness value. The fitness value determines which individual will move to the next community. The strength of an individual depends on his/her fitness value, and a good individual has a high fitness value if it is a maximization problem and a low fitness value if it is a minimization problem (**Table 1**) [2].

Genetic algorithms used in solving any problem consist of the following components:

1. Representation of individuals forming the community as a sequence ("chromosome").

2. Creation of the startup community.

3. Determining the eligibility of individuals and establishing the evaluation function.

4. Genetic operators for obtaining new populations.

5. Control variables (probabilities of crossover and mutation operators and the number of members in the ensemble).

Profile number	Profile section
1	IPE200
2	IPE220
3	IPE240
4	IPE270

Table 1.
Profile numbers and codes of these profiles.

3. Coding

The most important feature that distinguishes genetic algorithms from other methods is the use of codes representing these variables instead of variables. The first step in applying the genetic algorithm to any problem is to choose the most appropriate coding type for the problem. 2.4.1 binary coding.

3.1 Binary coding

In binary coding, each individual consists of the numbers 0 and 1 (**Table 2**). Chromosome length of the individual should be selected to include all points between the lower and upper limits of the parameter or variables. For a function with more than one variable, the chromosome length is equal to the sum of the chromosome lengths determined for each variable [1].

Although binary coding is a very common type of coding, it has some drawbacks. In multivariate function optimizations, the individual length obtained depending on the lower and upper limits of the variables is very large. Also traveling salesman scheduling etc. In complex optimization problems, binary coding does not fully represent the design space. For this reason, in addition to binary coding, real number coding and permutation coding types are also used.

3.2 Permutation coding

In permutation coding, the chromosome length is equal to the number of design variables. Permutation coding type is especially preferred in problems where design variables consist of more than one subvariable. The superiority of this coding type in shape optimization is illustrated with a simple example (Example 2.1.). In this coding type, the codes of the design variables consist of randomly selected numbers between 1 and the number of design variables (**Table 3**) (Example 1). In the problem of "optimizing sizing of the truss in **Figure 1**," the code of any individual to be included in the starting group is shown separately as permutation coding and binary coding.

Number	1. rod				2. rod				3. rod				4. rod				5. rod			
Prf number	1	2	3	4	1	2	3	4	1	2	3	4	1	2	3	4	1	2	3	4
Code	1	2	3	4	1	2	3	4	1	2	3	4	1	2	3	4	1	2	3	4

Table 2.
Code of a random individual in binary coding type.

	1.rod	2.rod	3.rod	4.rod	5. rod
Code	3	1	4	2	3

Table 3.
Code of a random individual in permutation coding type.

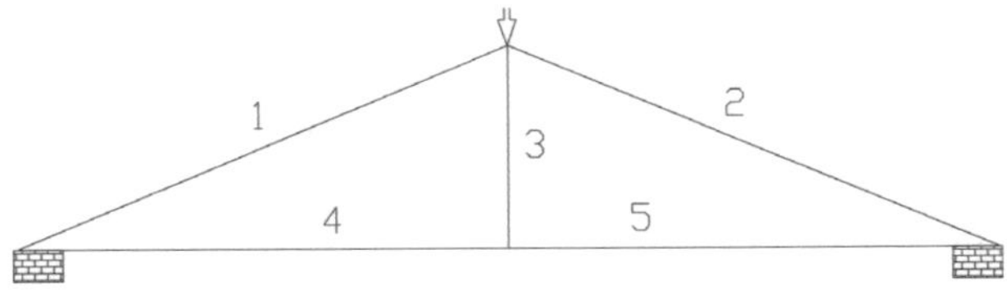

Figure 1.
Truss beam model is used to explain coding types.

4. Creating the initial population

Another important feature that distinguishes genetic algorithms from other methods is that it performs a search within a community of points, not point-to-point [1]. Therefore, the first step of the genetic algorithm is the creation of the initial population. In a problem with variable number L, it is recommended that the number of elements in the initial collection is between 2 L ~ and 4 L [3]. The initial population is simply randomly generated. However, especially in limited optimization problems, as a result of the random creation of the initial population, unsuitable solutions that do not meet the boundaries emerge. To eliminate this situation, it is to benefit from the heuristic methods developed for the problem. There are various studies in the literature that use heuristic methods in the creation of the initial community. For example, Grefenstette used gredgreedyristics in the traveling salesman problem, Kapsalis et al. used the minimum tree approach in the Steiner tree problem, Thiel and Vass used the addition-subtraction heuristics in the 0−1 backpack problem, and Chen et al. used the Campbell Dubek-Smith, and Dannenbrigt heuristics for the scheduling problem [4].

5. Evaluation

In each generation, the fitness values of the individuals in the community are calculated with an evaluation function. The fitness value plays a role in determining which candidate solutions from the existing community will be used to obtain new candidate solutions that will form the next community. The evaluation function used in genetic algorithms is the objective function of the problem. However, in some problems, objective functions may be limited by design variables. In this case, objective functions constrained by design variables should be converted to unconstrained objective functions independent of design variables with two transformations [5].

In the first transformation, the constrained objective function f(s) is transformed into the unconstrained objective function Ø(s) by applying the following transformation. Error functions are used in this transformation.

$$Ø(s) = \left[f(s) + R.\sum \Phi(Z) \right] \tag{1}$$

If $Z > 0$ ise $\Phi(Z) = Z2$

$$\text{If } Z \leq 0 \text{ ise } \Phi(Z) = 0 \tag{2}$$

In the second transformation, the unconstrained objective function Ø(s) is transformed into the fitness function F(s).

$$F(s) = \emptyset\ \max - \left[f(s) + R. \sum \Phi(gj(x) \right]\ F(s) = \emptyset\ \max - \emptyset(s) \tag{3}$$

f(s): Constrained objective function.
R: Predetermined error coefficient (R = 10,100 ...).
Φ(Z): Error function.
Ø(s): Unconstrained objective function.
Ø max: The maximum value of the unconstrained objective function.
F(s): Fitness function.

The fitness values of individuals are calculated from Eq. (3) [5]. Then, according to these fitness values, individuals to be used in the next stage are determined by any of the following methods.

When proportional selection methods are used in minimization problems, it is not possible to directly use the objective function to calculate the fitness values because these selection mechanisms are concerned with maximizing fitness. One method used to transform minimization problems into maximization problems is to multiply the objective function of the minimization problem by (−1). However, this method is not used in genetic algorithms due to the condition that the fitness values are positive. A minimization problem solved by genetic algorithms is transformed into a maximization problem by subtracting the objective function value of the problem from a large value determined at the beginning [1].

6. Selection methods

After the initial community is formed, in each generation, the members of the new community are selected from among the members of the existing community through a selection process. Individuals with a high fitness value have a high probability of acquiring new individuals. This operator performs natural selection artificially. The fitness of natural communities is determined by the ability of individuals to withstand the barriers of growth and reproduction. The various selection methods used are summarized below.

6.1 Proportional selection methods

6.1.1 Selection method with roulette wheel

The simplest selection method used in genetic algorithms is the Roulette Wheel method. In this method, the circle is divided into N equal spacing. In the circle spacings represents the elements. And the width of this interval is equal to the probability that this individual will be selected. In this case, the sum of the gap widths in the circle is equal to 1. During the selection phase, the circle is turned N times. In each turn, the individual indicated by the roulette dial is copied to the new community [1].

Although the roulette wheel method is simple, it has a stochastic error. In the new population, there is a difference between the expected number of copies of each individual and the number of copies that occur. In the next generations, this error increases and the search continues in many different directions than what was predicted theoretically. This event can cause generations to converge prematurely.

6.1.2 Stochastic ascending selection method

In this method, first of all, the expected copy number (mi) of the sequences in the ensemble is calculated. An integer copy of the expected value of each individual is registered in the new community. If the ensemble width is not reached, the fractional parts of the expected values are used with probability to fill the ensemble. There are two approaches to using fractional parts. The first approach is a choice without replacement. In this approach, the fractional part is taken into account as the probability value. For example, if the expected value of a copy of an individual is 1.5, a copy of this individual is included in the new community. The probability of the other copy being in the community is 50%. The other approach is choice by substitution. This approach uses a roulette wheel. The fractional parts are used to determine the intervals in the circle [4].

6.1.3 Stochastic universal selection method

This method is similar to the roulette wheel selection method. The most important difference is that the outer parts of the circle are divided into equal parts. The number of these parts is equal to the width of the ensemble. During the selection phase, the circle is rotated once. The copy number of an individual is determined by the number of pieces on the outside of the circle. In this case, the number of pieces falling on the weighted weight of an individual in the circle gives the number of copies of that individual [1].

6.1.4 Sequential selection method

The operation of this mechanism is summarized as follows. First of all, individuals in the community are ranked according to their fitness values from best to worst. The copy number is assigned to individuals with the help of a decreasing function. The most common assignment function used is linear. The copy numbers assigned with the help of a function are used to create the new community. At this stage, a new community is obtained by using one of the proportional selection mechanisms. It is stated that the number of copies of the best individual should be between 1 and 2 to control the election pressure in a community. The sequential selection mechanism has two major advantages. The first is that the scaling functions that should be used for the reasons explained do not need to be used in this selection mechanism. The second is to take the fitness values equal to the objective function values in minimization problems. This does not apply to proportional selection mechanisms.

6.1.5 Tournament selection method

In this mechanism, which has the advantages of the sequential selection mechanism, a group of individuals is randomly selected from the community. The individual with the best fitness value in this group is copied to the new community. This process is continued until the community width is reached. Generally, the group width is taken as two. However, it is also possible to increase this number [1].

6.1.6 Equilibrium selection method

The functioning of this mechanism is different from the others. In the methods described, firstly, a new population is obtained by selecting individuals from the existing population, and new individuals are obtained by applying genetic operators to the obtained population. The new individuals obtained are taken into the new community by replacing them with the individuals used to create them. In this method, genetic operators are applied to one or two individuals selected using the linear sequential selection method, and a new population is formed by replacing the new individuals obtained with the individuals with the lowest fitness value in the existing population.

6.1.7 Developed back-controlled selection operator (BCSO)

The selection process in existing selection operators is performed among individuals composing the population in the same generation. In these selection operators, the selection is carried out according to the fitness values of each individual. The fitness value of the individual in the existing generation is compared with the fitness value in the same generation. The BCSO differs from existing selection operators because the fitness value of the individual is compared with the fitness value in the previous generation. If the fitness value of the individual is more than the one in the preceding generation, this individual would keep their own position. Otherwise, if the fitness value of the same individual is less than or equal to the fitness value in the preceding generation, this individual would be discarded from the population. The individual would be copied to the population in the preceding generation to replace this individual.

7. Crossover operator

Crossover is a search operator that provides access to similar but unexplored regions of the search space by exchanging information between different solutions. With the crossover operator, two different individuals, which will give new points in the search space, are obtained by interchanging certain parts of two randomly selected individuals from the community. One-point, two-point, and uniform crossover methods are widely used in the literature.

7.1 One-point crossover operator

In this operator, the crossover point is randomly chosen between 1 and L-1. In two matched individuals, the segments after this cut-off point are replaced and two new individuals are obtained. In **Table 4**, a point crossover operator was applied to the parts of the individuals after the fifth site.

1	**0**	**1**	**1**	**0**	**1**	**0**	**1**	**0**	**1**	**1**	**1**	**0**	0	1	1
Parent 1												Child 1			
1	1	1	1	0	0	1	1	0	0	1	1	0	**1**	**0**	**1**
Parent 2												Child 2			

Table 4.
One-point crossover operator.

1	**0**	**1**	**1**	**0**	**1**	**0**	**1**	**1**	**0**	**1**	1	0	0	**0**	**1**
Parent 1								Child 1							
1	1	1	1	0	0	1	1	1	1	1	**1**	**0**	**1**	1	1
Parent 2								Child 2							

Table 5.
Two-point crossover operator.

0	1	1	0	0	1	0	0								
Temporary array															
1	**0**	**1**	**1**	**0**	**1**	**0**	**1**	**1**	0	0	**1**	**0**	0	**0**	**1**
Parent 1								Child 1							
1	0	0	1	0	0	1	0	1	**0**	**1**	1	0	**1**	1	1
Parent 2								Child 2							

Table 6.
Uniform crossover.

7.2 Two-point crossover operator

The only difference between these crossover operators from the one-point crossover operator is that two different intercept points are chosen randomly between 1 and L^{-1}. New individuals are obtained by relocating the regions of matched individuals between these breakpoints. An example of a two-point crossover performed at sites between the third and sixth cut points is given (**Table 5**).

7.3 Uniform crossover operator

In this operator, a random individual is temporarily created in binary order to be used in uniform crossover. The chromosome length of this individual is equal to the chromosome lengths of the other individuals in the community and n of the chromosome of the transient individual. If the code of the site is 0, the first new individual is n. The code of the 1st old individual on the site is the n of the second new individual. The code of the 2nd former individual will be used on the site, likewise, the chromosome of the temporary individual n. If the code of the site is 1, the n. 2. the old individual's code on the site is the n of the second new individual. The code of the 1st former individual will be used on the site. An example of uniform crossover is given in Ref. [6]. The fact that the number of temporary individuals to be used in this crossover method is equal to the number of pairs to be crossed over in the community and that these temporary individuals are not similar to each other is effective in the efficiency of the crossover operator (**Table 6**).

8. Mutation operator

When working on a limited population, there is a possibility that some genetic information in the population will be lost prematurely after a while. All the genes that

First-person before mutation							Second-person before mutation						
0	1	0	1	0	**1**	1	0	1	0	1	0	**0**	1
0	1	0	1	0	**0**	1	0	1	0	1	0	**1**	1
First-person after mutation							Second-person after mutation						

Table 7.
Mutation of two different individuals with binary code.

make up a chromosome can be the same. It is not possible to replace such a chromosome with the crossover operator. In this case, the gene in the randomly selected site on the chromosome is modified by interfering with the chromosomes that make up the community at a certain rate [1] (**Table 7**). It is recommended to take the mutation rate between 0.0001 and 0.05 [7]. We have the opportunity to produce individuals with high fitness values, which we cannot produce by crossover, by mutation. In addition, individuals with high fitness are at risk of being spoiled as a result of mutation [8].

9. Factors affecting the performance of genetic algorithm

Mutation rate: If the chromosomes start to resemble each other and are still far from the solution points, the mutation process is the only way to get the genetic algorithm out of its jam. However, giving a high value will prevent the genetic algorithm from reaching a stable point.

The number of crossover sites: Normally, crossover is performed at a single point, but research has shown that multipoint crossover is very useful in some problems.

How to evaluate the individuals obtained as a result of the crossover: It is sometimes important whether the two individuals obtained will be used at the same time.

Whether the generations are disjointed: Normally, each generation is created entirely dependent on the previous generation. In some cases, it may be useful to create the new generation with the old generation together with the individuals of the new generation obtained so far.

How the coding representation is made: However, it is not clear enough how it is a point that affects the performance of the genetic algorithm. Binary coding, permutation coding, and real number coding are the most common coding methods.

How the success evaluation is done: A poorly written evaluation function can prolong the runtime or cause the solution to never be reached.

10. Control parameters

In a genetic algorithm, there are three different control parameters: the size of the initial population, the crossover rate, and the mutation rate. The performance of a genetic algorithm is very dependent on the value of these parameters. Therefore, before applying genetic algorithms to a problem, the best combination of parameters must be determined.

10.1 The size of the starting group (N)

The size of the initial population is related to the convergence of the algorithm. The performance of the genetic algorithm decreases when the size of the initial population is small, because small heaps are insufficient to sample the search space and cause untimely convergence. The large initial population, on the other hand, increases the efficiency of the search, as it provides a very good sampling of the solution space and prevents premature convergence. However, when the size of the initial population is large, the evaluation of the sequences at each iteration increases the runtime and convergence occurs very slowly.

10.2 Crossover ratio

The crossover ratio controls the frequency of use of the crossover operator. In each new stack, the crossover operator is applied to Pç.N arrays. The high crossover rate realizes community variability quickly. A low-crossover rate causes the search to be very slow.

10.3 Mutation rate

Each element of each sequence in the new heap obtained by the selection process undergoes random changes with a probability equal to the Pm mutation rate. Since the high mutation rate turns the search made by the genetic algorithm into a random search, a low mutation rate should be preferred.

11. Conclusions

Genetic algorithms are one of the methods used as an alternative to traditional methods in solving optimization problems and are widely used in topology, shape, and size optimizations in the field of engineering. Genetic algorithms start with a population called the initial population, rather than a single individual that is likely to be the optimum solution to the solution of any optimization problem. In solving optimization problems with genetic algorithms, the risk of being caught in local optimum traps is less than in traditional optimization methods. Genetic algorithms use unconstrained objective functions in solving optimization problems. The use of unconstrained objective functions also enables the solution of multivariate optimization problems. Genetic algorithms, which are not an ordinary research technique, enable the discovery of new combination that comes with higher fitness values in each generation by using the best objective function and best fitness function values. Genetic algorithms can also be applied to optimization problems in cases where the design space is not discontinuous and convex, and where the design variables are mixed, continuous, and discontinuous. In classical optimization methods, it is necessary to find the first derivatives of the objective functions. In some complex optimization problems, finding the derivative of the objective function is either very difficult or impossible. In the use of genetic algorithms in optimization problems, there is no need for derivatives of objective functions, unlike classical optimization methods.

DOI: http://dx.doi.org/10.5772/intechopen.1000576

Author details

Mustafa Kaya
Aksaray University, Turkey

*Address all correspondence to: mkaya@aksaray.edu.tr

References

[1] Goldberg DE. Genetic Algorithms İn Search, Optimization and Machine Learning. Addison-Wesley Publishing Company, Inc; 1989

[2] Michalewicz Z. Genetic Algorithms +Data Structures=Evolutions Programs. Berlin Heidelberg, New York, USA: Springer-Verlag; 1996

[3] Turgut P. Determination of Live Load Combinosomes in Structures by Genetic Algorithms. Ankara, Turkey: Fırat University, Institute of Science and Technology; 1995

[4] Altiparmak F. Topological Optimization of Communication Networks with Genetic Algorithms. Ankara, Turkey: Gazi University Institute of Science and Technology; 1996

[5] Rao S. Engineering Optimization. Wiley Eastern Limited, Publishers And New Age İnternational Publishers, Ltd.; 1995

[6] Syswerda G. Uniform crossover in genetic algorithms. In: Third International Conference on Genetic Algorithms. 1989. pp. 2-9

[7] Erbatur F, ve Hasancebi O. Structural Design Optimization Using Genetic Algorithm, Department Of Civil Engineering Middle East Technical University, Uğur Ersoy symposium. 1999

[8] Holland JH. Adaptation in Natural and Artificial Systems. Ann Arbor: The University of Michigan Press; 1975

Chapter 9

On the Problems Facing Design Engineers and Methods for Their Solution in the Design of Reinforced Concrete Structures

Andrey Dolganov

Abstract

In general, this chapter is devoted to the optimization of reinforced concrete structures according to the criterion of minimizing their cost. All the design solutions discussed below do not contradict the established requirements of building regulations. It is shown that the current building codes and regulations do not fully disclose and quantitatively describe the reliability of the designed structures. In this chapter, an attempt was made to reveal the resources to reduce the cost of construction projects.

Keywords: reliability, probability of failure-free operation, plastic hinge, limit states, design resistance of concrete and reinforcement, standard deviation

1. Introduction

The goal of any project is to provide reliable and cost-effective functionality. Since I was asked to pay attention to the development of the design of reinforced concrete structures, this work is devoted to the analysis of the design problems of these structures.

Let us define the terms right away. So, by reliability we mean the probability of failure-free operation of building structures of buildings and structures. That is, we will give a quantitative assessment of the reliability of structures.

It is known that with insufficient reliability, there is a high risk of failure of construction projects. With excessive reliability, society incurs unreasonably high costs. That is, the issues of development and application in practice of the theory of reliability of building structures are related to socioeconomic issues.

Thus, we set ourselves the solution of the following problems.

1. Mathematical substantiation of the required level of reliability of building structures for structures of different levels of responsibility [1].

2. Justify the need to use probabilistic methods as the main ones in the design of building structures [2, 3].

3. Show the advantages of probabilistic calculation methods [4, 5].

The problem of assigning the optimal level of reliability has existed for a long time and, unfortunately, has not yet resolved. Design according to current building regulations (hereinafter referred to as BR for brevity) does not guarantee sufficient reliability, which is confirmed by regular construction accidents [2]. We can also state that design according to BR does not guarantee the equal reliability of sections of the same type of structures and elements. Often, calculations of reinforced concrete structures lead to both excessive and insufficient reliabilities. Existing standards regulate the reliability of building structures only to some extent. So, according to these documents, the reliability (security) of the standard resistances of materials is taken to be 0.95 (1.64σ), and the security of the design resistances approaches 0.99865 (3σ), because the standard resistances are divided into reliability factors for materials that are greater than 1.

Numerous other proposals for normalizing the failure probabilities $Q(t)$, in our opinion, are not sufficiently substantiated. For example, in [6], it is proposed to determine the probability of going beyond the boundaries of the limiting state:

$$Q(t) = 10^{-5}\xi_S T/L, \tag{1}$$

here ξ_S – factor of social significance (**Table 1**); T – regarded term of exploitation in years; L – an average amount of persons that are located inside building or nearby it for a regarded term when risk is estimated.

For buildings with a normal level of responsibility, the required reliability according to (1) will be:

$$1 - Q(t) = 10^{-5} \times 0.5 \times 50/50 = 0.999995 \text{ or } 0.955.$$

Firstly, the required value of reliability, as shown by the experience of designing, surveys of buildings and structures, as well as investigations of construction accidents, is excessive, and secondly, it is practically unattainable.

For example, let us determine the reliability for the structures of structures of a normal level of responsibility with known probabilities of the mechanical characteristics of materials and loads. To do this, we use the total probability formula (2). At the same time, we assume that the mechanical characteristics of materials and loads are independent and joint random variables, since the appearance of one random variable does not depend on the appearance of another.

$$P(A \cdot B) = 1 - \left[P\left(A^{/}\right) + P\left(B^{/}\right) - P\left(A^{/}\right)P\left(B^{/}\right)\right] \tag{2}$$

Type of buildings	ξ_S
The meeting place of people, dams	0.005
Living, office, trade and industrial buildings	0.05
Bridges	0.5
Towers, masts, structures at sea bottom	5

Table 1.
Factor of social significance ξ_S. *From [6].*

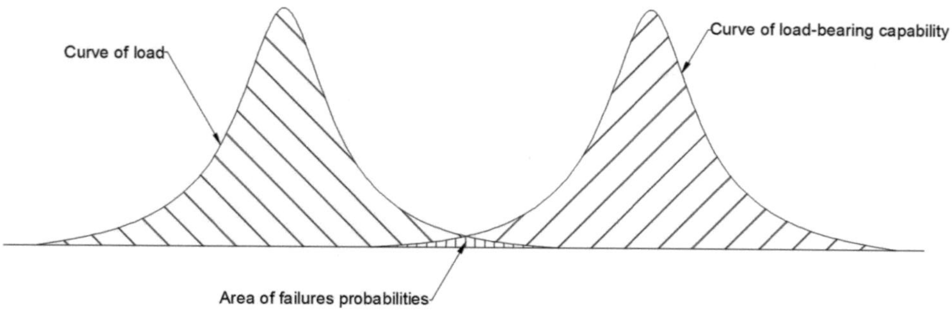

Figure 1.
For question about assignment of reliability level.

where $P(A')$ and $P(B')$ are probabilities of opposite events A (security of the mechanical characteristics of materials) and B (security of the design load):

$$P\left(A'\right) = 1 - P(A) = 1 - 0.99865 = 0.00135, P\left(B'\right) = 1 - P(B) = 1 - 0.95 = 0.05. \quad (3)$$

The expression $\left[P\left(A'\right) + P(B') - P\left(A'\right)P(B')\right]$ is the area of failure probabilities in **Figure 1**. Let us substitute the known values of the probabilities of random variables into formula (2) and determine $P(A{\cdot}B)$:

$$P(A \cdot B) = 1 - (0.05 + 0.05 - 0.05 \times 0.05) = 0.9025$$

or 1.3σ (for the normal distribution law).

With the reliability of the resistance of materials and loads of about 3σ, we have:

$$P(A \cdot B) = 1 - (0.00135 + 0.00135 - 0.00135 \times 0.00135) = 0.997$$

or 2.78σ (for the normal distribution law).

With the reliability of the resistance of materials and loads of about 4σ (0.999968), we have:

$$P(A \cdot B) = 1 - (0.000032 + 0.000032 - 0.000032 \times 0.000032) = 0.999936$$

or 3.83σ (for the normal distribution law).

Thus, the actual reliability of structures for structures of reduced, normal, and increased levels of responsibility corresponds to the values of $0.9025\,(1.3\sigma), 0.997\,(2.8\sigma)$ *and* $0.9^4 36\,(3.8\sigma)$.

Table 2 summarizes our proposals for assigning the level of reliability of building systems of different responsibility classes.

In principle, in our opinion, it would be more correct to determine any load, including snow load, using, for example, Pearson curves (**Table 3, Figure 2**):

$$\chi = \frac{r_3^2(r_4+3)^2}{4\left(4r_4 - 3r_3^2\right)\left(2r_4 - 3r_3^2 - 6\right)}, \quad (4)$$

where r_3, r_4 are the main statistical moments obtained by dividing the central statistical moments by the standard deviation of the statistical sample to the power of 3 and 4, respectively.

Level of building's responsibility	Probability of failure-free operation according to the design limit state[1]
High	0.9999 (4σ for the normal distribution law)
Conventional	0.9986 (3σ for the normal distribution law)
Low	0.9772 (2σ for the normal distribution law)

[1]*When the condition is met: the design parameter must be no more than the allowable one established by the building rules.*

Table 2.
Reliability of building systems for different classes of responsibility.

$\chi = -\infty$	$\chi < 0$	$\chi = 0$	$0 < \chi < 1$	$\chi = 1$	$\chi > 1$	$\chi = +\infty$
Type III	Type I	Type II, Type VII	Type IV	Type V	Type VI	Type III

Table 3.
Pearson distribution curves.

At $\chi = 0, r_3 = 0$ *and* $r_4 = 3$ – normal curve; $r_4 < 3$ type II; $r_4 > 3$ – type VII.

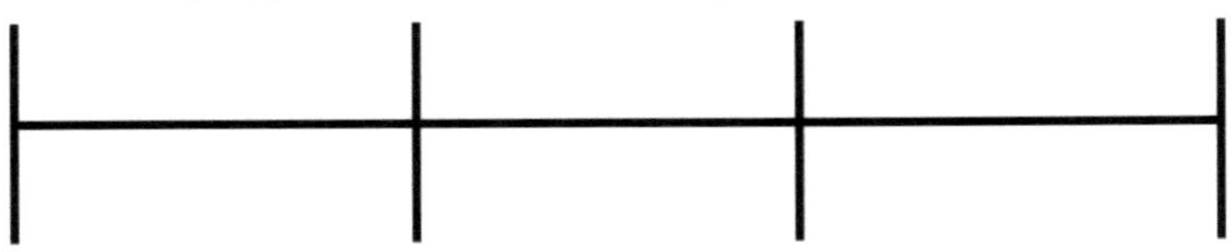

Figure 2.
To determine the type of distribution curve.

After we have determined the required level of reliability for building structures, we will evaluate their probability of failure-free operation.

Currently, a large number of methods for assessing the reliability of building structures are known.

In our opinion, the most universal is the method of statistical tests [7]. This method can be used to assess the reliability of both complex and simple building systems. Below is the algorithm for this method.

1. According to the accepted values of the parameters of known distributions, we assign t sets of realizations of random variables by the methods of statistical modeling: load $\tilde{F}$ and resistances of materials $\tilde{R}$. In this case, we use the Pearson distribution curves, which include, as mentioned above, the normal distribution law.

2. For each set of implementations, we perform calculations according to the formulas of current building rules. In this case, we determine the strength of the structural section $\tilde{H}_0$. After t calculations, we get t values of $\tilde{H}_0$.

3. We shall normalize the obtained calculation values of load-bearing capability $\tilde{H}_i = \tilde{F}/\tilde{H}_0$. Normalized values are approximated to 1. As result, the laboriousness of calculation is decreased.

4. We shall define density for distribution of random variables with assistance of smoothing the combination $\tilde{H}_i (i = 1, 2, \ldots, t)$ of one of Pirson's curves family, including also a normal curve.

5. We integrate the normalized curve up to the value of the given security (up to the values indicated in **Table 3**). We get the calculated parameter of the given security.

We assign the number of statistical tests from the condition of the minimum fluctuations of four statistical moments: 100,000.

2. Linearization method for assessing the reliability of reinforced concrete structures

This method is simpler, it assumes a formula dependence of the function of random arguments, and this dependence must be differentiable [7].

Let us evaluate the reliability of Section 17 (**Figure 3**) of a three-span $(I_1 = 6.178\ m, I_2 = I_3 = 5.300\ m)$ continuous reinforced concrete beam with a section of 300 × 600 mm constant along its length.

The beam is made of $B40$ class concrete. Beam reinforcement made of $A500$ class reinforcement. Tensile reinforcement resistance: R_s = 435,000 kN/m^2. The modulus of elasticity of concrete in the calculations was taken taking into account its creep: 11994000 kN/m^2; rebar elasticity modulus – 2E8 kN/m^2. Section number 17 (**Figure 3**) is reinforced from above and below symmetrically: according to 5d25. The beam is subjected to static permanent and temporary loads, as well as a movable wheel axial load of 98 kN. The diagram of the moments for the considered beam is shown in **Figure 3**.

Let us determine the utilization factor of the section:

$$\varphi(k) = M_q / M_0 \leq 1 \tag{5}$$

where M_q is the moment in the section from the design load, taken from **Figure 3**, $M_q = 510.43 kN \times m; M_0$ is the moment of internal forces, M_0 is determined by (6).

,
$$M_0 = \left[R_b b h_0{}^2 \alpha_m + R_{sc} A_s{}^{/} \left(h_0 - a^{/}\right)\right] \tag{6}$$

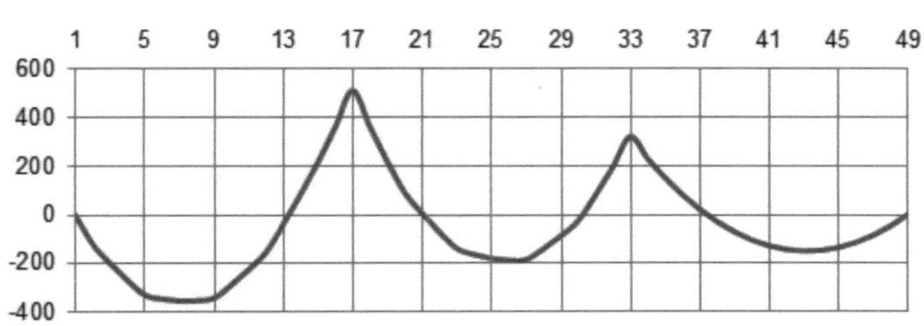

Figure 3.
Bending moment diagram, kN×m.

where R_b, R_{sc} are respectively the resistance of concrete and compressed reinforcement $R_b = 22400kN/m^2, R_{sc} = 400000kN/m^2; b, h_0$ are, respectively, the width and working height of the beam section $b = 0.30m, h_0 = 0.55m; a^/$ – structural size – distance from the center of gravity of the compressed reinforcement to the most compressed (least stretched) edge of the beam section, $a^/ = 0.05m; A_s^/$ – cross-sectional area of compressed reinforcement, $A_s^/(5\varnothing25) = 0.0024544m^2; \alpha_m$ – relative moment of internal forces, $\alpha_m = x(1 - 0.5\xi)$, ГДе$\xi = (R_s\mu - R_{sc}\mu^/)/R_b-$, where $\xi = (R_s\mu - R_{sc}\mu')/R_b$ – relative height of concrete compression zone; $\mu = A_s/bh_0; \mu^/ = A_s^/ /bh_0$.

For the convenience of differentiation, we multiply and divide the right side of expression (6) by bh_0^2, while assigning $k_{02} = bh_0^2$:

$$M_0 = \left(R_b{}^2\alpha_m + R_{sc}\mu^/ \cdot k_h\right) \times k_{02}, \tag{7}$$

where $k_h = (h_0 - a^/)/h_0$.

The initial data for probabilistic calculations are summarized in **Table 4**.

Further calculations are performed according to the following algorithm.

1. Let us take the partial derivatives in (5) with respect to R_b, R_s, R_{sc}, and q, and substitute the average values of the calculated parameters into the resulting expressions. When determining q, we take into account the correlation of the moment, the structural system and the nature of the loads:

$$m_q = q_n/q_c^{\cdot} \cdot M_{17}, \tag{8}$$

   where q_n, q_c are respectively, the normative (we take the average value) and the design load; $M_{17} = M_q = 510.43kN \times m$.

2. In the general case, according to formula (8), with average values of the calculated parameters, we determine the standard deviation σ_Y, in our case, the function $\varphi(k)$.

$$\sigma_Y^2 = \sum_{i=1}^{n}\left(\frac{\partial\phi}{\partial x_i}\right)_m^2 \sigma_{x_i}^2 + 2\sum_{i<j}^{n}\left(\frac{\partial\phi}{\partial x_i}\right)_m\left(\frac{\partial\phi}{\partial x_j}\right)_m r_{ij}\sigma_{x_i}\sigma_{xj} \tag{9}$$

$$r_{ij} = K_{ij}/(\sigma_i \cdot \sigma_j). \tag{10}$$

R_{bm}, kN/m^2	R_s, kN/m^2	R_{sc}, kN/m^2	m_q, $kN \times m$ (8)	A_s, m^2	$A_s^/$, m^2	b, m	h, m	a, m
37647.1	476538.5	438196.3	430.7	0.00245	0.00245	0.30	0.60	0.05
V_b	V_s	V_{sc}	V_q	μ	$\mu^/$	λ	k_c	k_h
0.135	0.053	0.053	0.106	0.01487	0.01487	10.3	0.658	0.909

Table 4.
Input data for statistical modeling.

where r_{ij}, σ_{x_i}, σ_{x_j} are the correlation coefficient and standard deviations of random variables X_i, X_j; K_{ij} is the correlation moment of random variables X_i, X_j; for independent random variables $r_{ij} = 0$.

3. For the section under consideration, we determine the coefficient k with a given probability:

$$\mathrm{k(t)} = 1 - \mathrm{t} \times \sigma_{\mathrm{Y}}, \tag{11}$$

where t is taken from **Table 2** depending on the responsibility of the building or structure.

In (5) we take partial derivatives with respect to $\mathrm{R_b}$, $\mathrm{R_s}$, $\mathrm{R_{sc}}$ and $\mathrm{m_q}$:

$$\partial\varphi(\mathrm{k})/\partial\mathrm{R_b} = -0.5\mathrm{m_q}\left(\mathrm{R_s}\mu - \mathrm{R_{sc}}\,\mu'/\mathrm{k_c}\right)^2/ \\ \left\{\mathrm{R_b}^2\mathrm{k_{02}}\left[\left(\left(0.5\mathrm{R_s}\mu - \mathrm{R_{sc}}\,\mu'/\mathrm{k_c}\right)/\mathrm{R_b} - 1\right)\left(\mathrm{R_s}\mu - \mathrm{R_{sc}}\,\mu'/\mathrm{k_c}\right) - \mathrm{R_{sc}}\mu'/\mathrm{k_c}\mathrm{k_h}\right]^2\right\}; \tag{12}$$

$$\partial\varphi(\mathrm{k})/\partial\mathrm{R_s} = \mathrm{m_q}\left[\mu\left(\left(0.5\mathrm{R_s}\mu - 0.5\mathrm{R_{sc}}\mu'/\mathrm{k_c}\right)/\mathrm{R_b} - 1\right) + 0.5\mu\left(\mathrm{R_s}\mu - \mathrm{R_{sc}}\mu'/\mathrm{k_c}\right)/\mathrm{R_b}\right]/ \\ \left\{\mathrm{k_{02}}\left[\left(\left(0.5\mathrm{R_s}\mu - 0.5\mathrm{R_{sc}}\mu'/\mathrm{k_c}\right)/\mathrm{R_b} - 1\right)\left(\mathrm{R_s}\mu - \mathrm{R_{sc}}\,\mu'/\mathrm{k_c}\right) - \mathrm{R_{sc}}\mu'/\mathrm{k_c}\mathrm{k_h}\right]^2\right\}; \tag{13}$$

$$\partial\varphi(\mathrm{k})/\partial\mathrm{R_{sc}} = -\mathrm{m_q}\left[\mathrm{k_c}\mathrm{k_h}\mu' + \mathrm{k_c}\mu'\left(\left(0.5\mathrm{R_s}\mu - 0.5\mathrm{R_{sc}}\,\mu'/\mathrm{k_c}\right)/\mathrm{R_b} - 1\right)\right. \\ \left. + 0.5\mathrm{k_c}\,\mu'\left(\mathrm{R_s}\mu - \mathrm{R_{sc}}\mu'/\mathrm{k_c}\right)/\mathrm{R_b}\right]/ \\ \left\{\mathrm{k_{02}}\left[\left(\left(0.5\mathrm{R_s}\mu - 0.5\mathrm{R_{sc}}\mu'/\mathrm{k_c}\right)/\mathrm{R_b} - 1\right)\left(\mathrm{R_s}\mu - \mathrm{R_{sc}}\,\mu'/\mathrm{k_c}\right) - \mathrm{R_{sc}}\,\mu'/\mathrm{k_c}\mathrm{k_h}\right]^2\right\}; \tag{14}$$

$$\partial\varphi(\mathrm{k})/\partial\mathrm{m_q} = -1/\left\{\mathrm{k_{02}}\left[\left(\left(0.5\mathrm{R_s}\mu - 0.5\mathrm{R_{sc}}\,\mu'/\mathrm{k_c}\right)/\mathrm{R_b} - 1\right)\left(\mathrm{R_s}\mu - \mathrm{R_{sc}}\,\mu'/\mathrm{k_c}\right) - \mathrm{R_{sc}}\,\mu'/\mathrm{k_c}\mathrm{k_h}\right]\right\}. \tag{15}$$

The calculation results are shown in **Table 5**.

$\partial\varphi(\mathrm{k})/\partial\mathrm{R_b}$	$\partial\varphi(\mathrm{k})/\partial\mathrm{R_s}$	$\partial\varphi(\mathrm{k})/\partial\mathrm{R_{sc}}$	$\partial\varphi(\mathrm{k})/\partial\mathrm{q}$	$\sigma_{\mathrm{Rb}} = \mathrm{V_b R_{bm}}$	$\sigma_{\mathrm{Rs}} = \mathrm{V_s R_s}$	$\sigma_{\mathrm{Rsc}} = \mathrm{V_s R_{sc}}$	$\sigma_{\mathrm{q}} = \mathrm{V_q m_q}$
−3.01302 E-07	−3.5E-07	4.07474 E-09	0.000241232	5082.35	25328.37	23290.46	45.57
σ_{Y}, kN × m	$\lvert\mathrm{k}\rvert_{17}$	t_k	$P(t_k)$	$k_{0,95}$	$k_{0,99,865}$	q_n / q_c	k_{02}, m^3
0.01413	0.942	4.08	0.9^48	0.977	0.958	0.844	0.091

Table 5.
Reliability assessment by linearization method.

2.1 Conclusion

The reliability of section number 17 (**Figure 3**) in the considered beam is excessive. In this case, it is possible to reduce the reinforcement or increase, for example, the

σ_Y, kN × m	$\|k\|_{17}$	t_k	$P(t_k)$	$k_{0.95}$	$k_{0,99,865}$	q_n / q_c	k_{02}, m^3
0.01509536	0.977	1.52	0.93513	0.975	0.955	0.844	0.091

Table 6.
Results of reinforcement optimization of a reinforced concrete beam.

technological load. But here it is important to understand that the coefficient k must always be less than 1, while the building rules allow $k = 1$. Particularly sensitive to the randomness of the design parameters are metal structures. As shown by special calculations and the experience of investigating construction accidents, in such structures the coefficient k with a probability of 3σ (0.99865) in many cases has values less than 0.9. This fact can be explained in the following way. Reinforced concrete is a composite material. When in the section we have conditionally "bad" concrete, then the section is helped by a "good" material – reinforcement, and vice versa. The probability that there will be "bad" concrete and reinforcement in the section at the same time is much lower than if it were one material. Therefore, initially reinforced concrete structures are more reliable than, for example, metal ones.

Let us reduce the reinforcement: let us take $4d25 + 2d16$, $A_s = A_s^{/} = 0.0023656\ m^2$ (savings in the area of reinforcement 6.9% in one section of the beam) (savings in the reinforcement area of 6.9% in one section of the beam). Then the calculated indicators change (see **Table 6**).

Next, we will try to take into account the effect of static indeterminacy on the reliability of structures.

3. Structural reliability analysis of a three-span beam

The destruction of the technical system occurs under the condition of the formation of a set of a certain number of plastic hinges, highlighted in **Figure 4** shaded circles [7]. Let us compose the reliability structure of a three-span continuous beam. To do this, we determine the "weight," "significance," and "contribution" of the design sections in the reliability structure of the considered technical system.

Let us consider all possible schemes of destruction of the beam (**Figure 4**). Thus, the beam will collapse according to the first scheme if plastic hinges are formed in sections 1 and 2 (**Figure 4b**). The beam will fail according to the second scheme if plastic hinges are formed in sections 4 and 5 (**Figure 4c**). And so on.

Schemes of destruction of a technical system are disjoint events. Let us pass from an arbitrary function of the algebra of logic to an orthogonal disjunctive-normal matrix form:

$$f(x_1, \dots, x_n) = \begin{vmatrix} K_1 \\ K_2 \\ K_3 \\ \vdots \\ K_m \end{vmatrix} = \begin{vmatrix} K_1 \\ K_1^{/} K_2 \\ K_1^{/} K_2^{/} K_3 \\ \vdots \\ K_1^{/} K_2^{/} K_3^{/} \dots K_{m-1}^{/} K_m \end{vmatrix}, \tag{16}$$

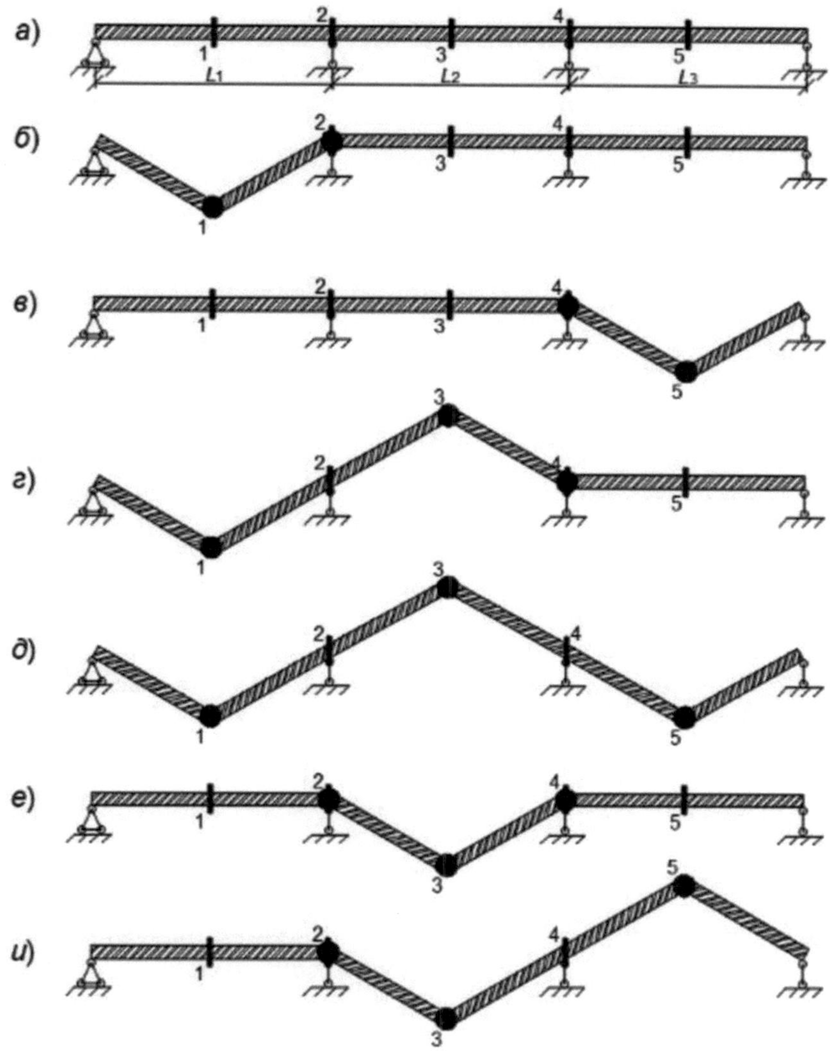

Figure 4.
Schemes of destruction of a three-span continuous beam.

where $K_i, i = 1, \ldots, 6$ is a conjunction describing the destruction scheme i; $K_i^{/}, i = 1, \ldots, 5$ is negation of conjunction *i*.

For the fracture schemes shown in **Figure 4**, we have the following conjunctions:

$$K_1 = x_1 \cap x_2; K_2 = x_4 \cap x_5; K_3 = x_1 \cap x_3 \cap x_4; K_4 = x_1 \cap x_3 \cap x_5; K_5 = x_2 \cap x_3 \cap x_4; K_6 = x_2 \cap x_3 \cap x_5. \quad (17)$$

Expression (17) can be simplified.

$K_1 = x_1x_2; K_2 = x_4x_5; K_3 = x_1x_3x_4; K_4 = x_1x_3x_5; K_5 = x_2x_3x_4; K_6 = x_2x_3x_5.$

Let us write the negations of conjunctions $K_i^{/}, i = 1, \ldots, 5$:

$$K_1^{/} = x_1^{/} \cup x_1 \cap x_2^{/}; K_2^{/} = x_4^{/} \cup x_4 \cap x_5^{/}; K_3^{/} = x_1^{/} \cup x_1 \cap x_3^{/} \cup x_1\ x_3\ x_4^{/};$$
$$K_4^{/} = x_1^{/} \cup x_1 \cap x_3^{/} \cup x_1 \cap x_3 \cap x_5^{/}; K_5^{/} = x_2^{/} \cup x_2 \cap x_3^{/} \cup x_2 \cap x_3 \cap x_4^{/}; \quad (18)$$
$$K_5 = x_2^{/} \cup x_2 \cap x_3^{/} \cup x_2 \cap x_3 \cap x_5^{/}.$$

Expression (18) will also be simplified.

$$K_1^{/}\ K_2 = x_1^{/} x_4 x_5 + x_1 x_2^{/} x_4 x_5; K_1^{/} K_2^{/} K_3 = x_1 x_2^{/} x_3 x_4 x_5^{/};$$
$$K_1^{/} K_2^{/} K_3^{/} K_4 = x_1 x_2^{/} x_3 x_4^{/} x_5; K_1^{/} K_2^{/} K_3^{/} K_4^{/} K_5 = x_1^{/} x_2 x_3 x_4 x_5^{/}; \quad (19)$$
$$K_1^{/} K_2^{/} K_3^{/} K_4^{/} K_5^{/} K_6 = x_1^{/} x_2 x_3 x_4^{/} x_5.$$

Let us substitute into (16) combinations of conjunctions from (19):

$$f(x_1, \dots, x_6) = x_1 x_2 + x_1^{/} x_4 x_5 + x_1 x_2^{/} x_4 x_5 + x_1 x_2^{/} x_3 x_4 x_5^{/} + x_1 x_2^{/} x_3 x_4^{/} x_5 + x_1^{/} x_2 x_3 x_4 x_5^{/} + x_1^{/} x_2 x_3 x_4^{/} x_5. \quad (20)$$

From (20) we determine the probability of failure or the reliability of the functioning of the kinematic mechanisms) of the considered beam. In this case, the reliability of sections 1, …, 5 (**Figure 4**) is assumed to be the same, indicated in **Table 6**, $x_1^{/} = x_2^{/} = x_3^{/} = x_4^{/} = x_5^{/} = 0.93513$.

Then the probability of formation of a plastic hinge in the calculated section will also be the same: $x_1 = x_2 = x_3 = x_4 = x_5 = 1 - 0.93513 = 0.06487$.

Substituting the known design parameters in (20) we obtain.

$$R_{sys} = 1 - f(x_1, \dots, x_5) = 1 - 0.00935 = 0.99065 \text{ or } 2,35\sigma. \quad (21)$$

Thus, the reliability of the considered beam is 1.55 times greater than the reliability of its design sections: $2.35\sigma/1.52\sigma = 1.55$.

Let us change the reliability of the sections so that the reliability of the beam as a technical system is equal to the value indicated in **Table 2** (0.9986). To do this, we use special quantitative parameters that characterize the structure of the system reliability: "weight," "significance," and "contribution."

The formulas for determining the indicated reliability parameters are given in their final form without conclusions.

Let us determine the "weight" of the element x_i in a system consisting of n elements or the ratio of the "weight" of the Boolean difference in the argument x_i to the number of all sets of the n-dimensional logical space:

$$g_{x_1} = \frac{G\{\Delta_{x_i} y(x_1, \dots, x_n)\}}{2^n}. = \sum_j 2^{-(r_j - 1)} + \sum_f 2^{-(r_f - 1)}, \quad (22)$$

where $f = 1, \dots, k; j = 1, \dots, l; q = 1, \dots, p; k + l + p = m; r_f, r_j, r_q$ are the ranks of elementary conjunctions; k, l, p – is number of conjunctions containing $x_i^{/}, x_i$ and not containing the i-th argument; m is the total number of conjunctions in the original logic algebra function represented in orthogonal disjunctive-normal form; n is the number of independent variables of the original function.

The "weight" of the Boolean difference (22) characterizes the importance of the element x_i for the reliability of the system. The values of the normalized or relative "weight" for each section of the considered system are given in **Table 6**.

The rate of change in reliability in the design sections of the technical system will be determined using the concept of "significance." The significance of the element x_i in the system $y(x_1, \dots, x_n)$ is a partial derivative of the probability of failure-free operation of the system R_c by the probability of failure-free operation of the element R_i, i.e.,

$$\zeta_{x_i} = \frac{\partial P\{y(x_1, \dots, x_n) = 1\}}{\partial P\{x_i = 1\}} = \frac{\partial y_c}{\partial x_i} \tag{23}$$

The criterion "significance" of the element allows you to determine the elements that provide the maximum increase in the reliability of the entire system. The significance of the design cross sections of the system under consideration is also given in **Table 6**.

The product of the probability of non-failure operation R_i of the element x_i and its "significance" in the system $y(x_1, \dots, x_n)$is called the "contribution"

$$B_{x_i} = R_i \frac{\partial R_c}{\partial R_i} \tag{24}$$

or "specific contribution"

$$b_{x_i} = B_{x_i} / \sum_{i=1}^{n} B_{x_i} \tag{25}$$

The criterion "contribution" of the element B_{xi} characterizes the location of the element x_i in the structure of the system $y(x_1, \dots, x_n)$, the conditions of its operation and the connection with the probability of failure-free operation of all n elements of this system, including the i-th element. In addition, the "contribution" criterion makes it possible to rationally determine the order of restoring elements in a technical system.

The values of "specific contributions" of the calculated cross sections of the system under consideration are also given in **Table 7**.

The introduced differential characteristics of the elements $g_{x_i}, \xi_{x_i}, b_{x_i}$ make it possible to visually see the distribution of the role of all elements on a given reliability structure in solving a specific problem.

Reliability parameter	**Estimated cross section *i***				
	1	**2**	**3**	**4**	**5**
g_i	0.375	0.375	0.250	0.375	0.375
g_i/g_{max}	1.00	1.00	0.67	1.00	1.00
ξ_i	0.07145	0.07145	0.01472	0.07145	0.07145
ξ_i/ξ_{max}	1.00	1.00	0.21	1.00	1.00
b_i	0.238	0.238	0.049	0.238	0.238
b_i/b_{max}	1.00	1.00	0.21	1.00	1.00

Table 7.
Analysis of the reliability structure of a continuous beam.

For example, by adjusting the reliability of sections 1, 2, 4, 5 (**Figure 4**), we will achieve that the reliability of the system becomes 0.9986. Given the equal importance, significance and contribution of sections 1, 2, 4, 5 to the reliability structure, we will increase the reliability of these sections by 0.975. The probability of failure-free operation of Section 3 is not changed and remains the same 0.93513. Substituting the known parameters into (21), we obtain the reliability of the system equal to 0.9986.

3.1 Conclusion

The concepts of "weight," "significance," and "contribution" of elements to the reliability structure of a technical system, introduced into the design practice, will allow optimizing the design, linking operating costs with the costs associated with reconstruction, with overhaul of the most important elements.

Thus, the design of building systems, taking into account their reliability, can reduce the risk of construction accidents in some cases, and reduce the cost of the facility in other cases.

4. Accounting for parallel operation of reinforcement in sections of reinforced concrete structures

Below we will consider the possibility of saving reinforcement in reinforced concrete structures with a number of reinforcing bars of at least 5. The idea is simple: in building codes and rules, the resistance of reinforcement depends on the class of reinforcement and does not depend on the number of reinforcing bars in the section of the structure. At the same time, it is clear that with an increase in the number of reinforcement bars in the structure section, the probability of simultaneous failure of all bars decreases.

The resistances of materials are random variables. The design of structures is carried out according to an initially unfavorable scenario from the point of view of strength. Therefore, if we use the main provisions from probability theory and mathematical statistics in the calculation prerequisites, then the so-called effect of taking into account the parallel operation of the reinforcement in the section of the reinforced concrete structure is obtained.

The following conditions are fundamental when constructing a mathematical model for carrying out probabilistic calculations: The provision of resistances is 0.95; the calculated resistances are obtained by dividing the regulatory values by the reliability coefficient of the reinforcement. The calculated resistances obtained as a result of statistical tests are compared with the value specified in the building design standards for the considered class of reinforcement.

This chapter does not raise the issues of the required security of design resistances. Also, the problem of quantifying the reliability of structures of buildings and structures of different levels of responsibility is not considered. However, taking into account the effect of parallel operation of the valve in the design will allow for a more qualitative design of the design of a given reliability.

So, with an increase in the number of rebars, the probability of simultaneous failure of all the rebars decreases. This factor is not taken into account in the building rules. For example, for A500 reinforcement, the standard resistance is 500 *MPa*, and the calculated resistance is 435 *MPa* and does not depend on the number of rods.

At the same time, for each shipment of reinforcement, a quality certificate is attached, which indicates the melting number of reinforcing steel, the standard deviation and the average strength of the reinforcement in the batch. Therefore, for several batches, we can independently determine the inter-batching variation in the strength of reinforcement.

Let us simulate the process described above. In this case, the number of statistical tests will be set to 100,000 (with an infinite number of statistical tests, any distribution tends to normal). In the section of the structure, the reinforcement of the same diameter will be considered from one melt batch. It is clear that rebar of the same diameter can belong to different heat batches if it was rolled at different times at the same metallurgical plant or rolled at the same time, but at different metallurgical plants. Therefore, we take this decision in reserve.

1. With the normative design resistances of the reinforcement R_s and the known number and diameters of the reinforcement of each melting batch, we determine the conditional load-bearing capacity of the structure section N_C. For a section reinforced with bars, for example, two types of diameters (k = 2), but one class of reinforcement, the value of N_C will be equal to:

$$N_C = R_s \times \pi \times \left(n_1 d_1{}^2 + n_2 d_2{}^2\right)/4, \tag{26}$$

   where n_1, n_2, d_1 and d_2 is respectively, the number and diameter of the first and second batches of melts.

2. With the known inter-batch variation coefficient V_{MP} for the reinforcement of each batch-melting, we set the realizations of the reinforcement resistances, which will be the average values of $\tilde{R}_{sm}$ for statistical tests of the resistances of each batch-cast.

3. Let us carry out statistical tests for the reinforcement of each batch-cast: we will obtain the realizations $\tilde{R}_{si}$.

4. Let us determine the conditional load-bearing capacity of the section for random realizations of reinforcement resistances.

$$\tilde{N} = \tilde{R}_{si} \times \pi \times \left(n_1 d_1{}^2 + n_2 d_2{}^2\right)/4 \tag{27}$$

5. Let us calculate the statistical characteristics of the conditional bearing capacity of the section (average value $\overline{N}$ and standard deviation σ_N).

6. Let us determine the conditional bearing capacity of the section with a security of 0.95 $\tilde{N}_{0.95}$ and divide it by the known reinforcement safety factor γ_s. Get N_{cal}:

$$N_{cal} = \left(\overline{N}\text{-}1.64\sigma_N\right)/\gamma_s \tag{28}$$

7. Let us determine the coefficient of parallel operation of the steel reinforcing k_{ps} through the ratio of N_{cal} to N_C N_{cal} to N_C $\left(k_{ps} = N_{cal}/N_C\right)$. The coefficient k_{ps} means

a safety margin and depends on the batch and inter-batch coefficients of variation in the strength of the reinforcement, as well as on the number of reinforcing bars in the section of the reinforced concrete structure of different batches-melts.

Example. Consider reinforced concrete columns during the reconstruction of the Olympic sports complex in Moscow. According to the album 821–2.3-18-KP2, columns with a section of 500 × 500 are reinforced with 12∅25A500; 800 × 800 – 16∅25A500; 1000 × 1000 – 24∅25A500; 1200 × 1200 – 32∅25A500.

Let us perform statistical modeling of resistances according to the algorithm of paragraphs 1–7. The results of statistical modeling calculations for the above, as well as some other columns of the sports complex "Olympic" are summarized in **Table 8** and shown in **Figure 5**.

Reinforcement type	k_s	n_1	d_1, *mm*	n_2	d_2, *mm*	V_1	V_2	V_{mp1}	V_{mp2}	R_{sm1}, *MPa*	R_{sm2}, *MPa*
1	2	3	4	5	6	7	8	9	10	11	12
1	1.123	12	40	0	0	0.027	0.000	0.009	0.000	573	0
2	1.117	12	32	0	0	0.018	0.000	0.015	0.000	574	0
3	1.134	12	25	0	0	0.023	0.000	0.026	0.000	593	0
4	1.148	12	20	0	0	0.020	0.000	0.015	0.000	590	0
5	1.073	12	16	0	0	0.021	0.023	0.047	0.000	582	0
6	1.153	12	12	0	0	0.018	0.023	0.015	0.000	592	0
7	1.145	18	25	16	20	0.023	0.023	0.009	0.009	580	580
8	1.145	16	25	16	20	0.023	0.023	0.009	0.009	580	580
9	1.145	16	25	16	22	0.023	0.023	0.009	0.009	580	580
10	1.145	16	25	14	22	0.023	0.023	0.009	0.009	580	580
11	1.117	5	40	0	0	0.027	0.000	0.009	0.000	573	0
12	1.114	5	32	0	0	0.018	0.000	0.015	0.000	574	0
13	1.132	5	25	0	0	0.023	0.000	0.026	0.000	593	0
14	1.146	5	20	0	0	0.020	0.000	0.015	0.000	590	0
15	1.072	5	16	0	0	0.021	0.023	0.047	0.000	582	0
16	1.150	5	12	0	0	0.018	0.023	0.015	0.000	592	0
17	1.141	5	25	5	20	0.023	0.023	0.009	0.009	580	580
18	1.143	8	25	8	20	0.023	0.023	0.009	0.009	580	580
19	1.141	5	25	5	22	0.023	0.023	0.009	0.009	580	580
20	1.143	8	25	8	22	0.023	0.023	0.009	0.009	580	580
21	1.134	12	25	0	0	0.023	0.000	0.026	0.000	593	0
22	1.135	16	25	0	0	0.023	0.000	0.026	0.000	593	0
23	1.135	24	25	0	0	0.023	0.000	0.026	0.000	593	0
24	1.135	32	25	0	0	0.023	0.000	0.026	0.000	593	0

Table 8.
Accounting for parallel operation of reinforcement in a section of a reinforced concrete structure.

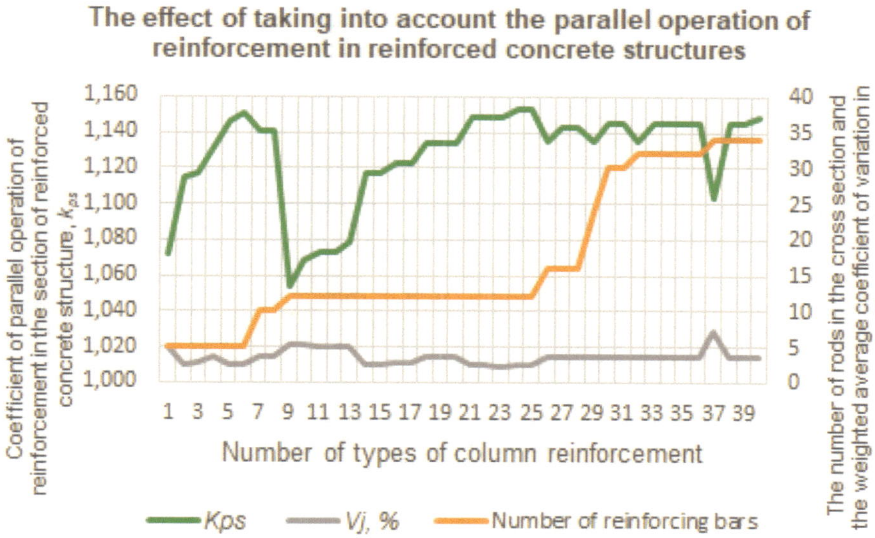

Figure 5.
The effect of the parallel operation of the armature depending on various factors.

As can be seen in **Figure 5**, the coefficient k_{ps} is significantly affected by the batch and inter-batch variation coefficients $V_1, V_2, V_{mp1}, V_{mp2}$, the number of reinforcing bars n_1, n_2, is somewhat less influenced, and the diameters of the reinforcing bars d_1, d_2 have no effect.

On **Figure 5** V_j is the geometric mean coefficient of variation:

$$V_i = \left(V_1{}^2 + V_{1m}{}^2 + V_2{}^2 + V_{2m}{}^2\right)^{1/2}, \tag{29}$$

where V_1, V_2, V_{1m}, V_{2m} are respectively, the coefficients of variation of batches of melts 1 and 2, inter-batch coefficients of variation of batches of melts 1 and 2.

4.1 Conclusions

When taking into account the effect of parallel operation of reinforcement in the section of a reinforced concrete structure, we obtain the following advantages.

1. It is possible to reduce the cross-sectional area of the longitudinal reinforcement by a factor of k_{ps}. Given the known statistical parameters (mean values and standard deviations of the reinforcement strength in heat batches, which we accept from quality certificates accompanying deliveries) and the number of reinforcing bars in the section of the elements, we multiply the design resistance of the reinforcement by the k_{ps} coefficient. Further, according to **Table 7**, we correct the value of the reinforcement area required for strength.

2. The use of the k_{ps} coefficient makes it possible to obtain a more accurate assessment of the probability of failure-free operation of the structure (the value of its reliability) and the building as a whole. With excessive reliability, we save resources or increase the allowable load on the structure. In case of insufficient

reliability of the structure, we can take timely measures to prevent the structure from going beyond the limits of the limit state.

3. Since the size of the overlap in the construction rules depends on the ratio of the calculated and actual areas of the reinforcement, we can reduce the length of the overlap by k_{ps} times. If we take into account that the reinforcement joints at the facility make up 15% of the total amount of reinforcement, then the metal savings will be 15% × 0.13 = 1.95% per one lap joint.

Thus, as shown by special calculations, within the framework of the construction of only one object, the use of probabilistic methods in the design can reduce the consumption of reinforcement by about 10%.

5. Benefits of using reinforced concrete beams in design

Consider an example. Welded beams made of C345 steel are connected to the reinforced concrete deck using stud bolts. The flooring concrete corresponds to class B30. The deck plate has a thickness of 130 *mm*. Profiled flooring H57–750-0.8 is used as fixed formwork.

The initial data for special calculations were also the results of surveys and tests of steel beams performed by TsNIISK them. V.A. Kucherenko, structural sections of the architectural and construction project 738.16.1-KR1.4 and 738.16.2-KR2.4. Calculations were performed for cases without and taking into account the operation of a monolithic reinforced concrete floor slab.

When taking into account the contribution of reinforced concrete flooring to the bearing capacity and rigidity of the floor, the concepts of reduced thickness and reduced modulus of elasticity (deformation) were used. Thus, the reduced thickness of the reinforced concrete slab was 10.9 *cm*.

The reduced modulus of elasticity of the reinforced concrete beam was determined by the formula (30):

$$E_{red} = (E_1 J_1 + E_2 J_2)/(J_1 + J_2), \tag{30}$$

where E_1 and E_2, J_1 and J_2 there are, respectively, the moduli of elasticity of the steel beam and the floor concrete, the moments of inertia of the steel beam and the reinforced concrete floor slab.

When determining J_2, the effective width of the reinforced concrete deck was taken into account in accordance with the building codes:

$$b_f^{/} = \left(2 \times 6h_f^{/} + b\right) \leq B \tag{31}$$

where $b_f^{/}$ there is an estimated width of reinforced concrete flooring; $h_f^{/}$ is the reduced thickness of the reinforced concrete flooring; b is the width of the steel beam flange; B is the step of the beams.

In the calculation model, in order to save metal, the reinforcement plates were not brought to the beam support by $10L_k$, where L_k is the length of one finite element. Loads were set according to the project with the codes indicated above. The normative and design loads are respectively equal to 8.08 and 9.08 kN/m^2.

Design load parameters for probabilistic calculations were determined: $q_m = 8.08$ кН/м^2, $\sigma_q = (9.08 - 8.08)/3 = 0.33$ кН/м^2, $V_q = 0.33\ /\ 8.08 = 0.041$. The design parameters of the mechanical characteristics of materials were determined at the coefficients of variation for steel – 0.04 and for concrete – 0.135. Average resistance of steel $R_{ym} = 345\,000/(1 - 1.64 \times 0.04) = 369\,221$ kN/m^2. The standard deviation of steel resistance is equal, kN/m^2 : $\sigma_s = 267\,550 \times 0.04 = 10\,702$.

For a uniformly distributed load, the function of the section utilization factor has the form.

$$\varphi(k) = \left(ql^2/8\right)/\left(R_y\gamma_c w\right), \tag{32}$$

where q is the value of linear load, *kN/m*; l is the design span of the beam; R_y is the resistance of steel; γ_c is coefficient of working conditions of the structure, $\gamma_c = 0.9$; w is modulus of section, m^3.

The standard deviation of the function $\varphi(k)$ is determined by (9). In the partial derivatives with respect to R_y and q in (9), instead of R_y and q, we substitute their average values R_{ym} and q_m. When calculating according to formulas (33)–(34), the coefficient γ_c is taken equal to 1; we reduce the load distributed over 1 m^2 to the linear load, multiplying it by the step of the beams B.

$$\partial\varphi(k)/\partial R_y = -q_m\, l^2/\left(8R_{ym}{}^2\,\gamma_c\, w\right), \tag{33}$$

$$\partial\varphi(k)/\partial q_m = l^2/\left(8R_{ym}\,\gamma_c\, w\right); \tag{34}$$

$$\partial\varphi(f)/\partial q_m = 5/384 \times l^4 \times B/(EJ); \tag{35}$$

$$\partial\varphi(f)/\partial E = -5/384 \times q \times l^4 \times B/\left(E^2J\right). \tag{36}$$

For loads distributed unevenly, differentiation was carried out numerically.

To compare the test results with probabilistic calculations, four types of beams were selected: B-12, B-45, B-49 and B-61. Selected calculation results are presented in **Tables 9** and **10**. For normal operation of structures, it is required that, with a design live load of 2.4 *kN/m*^{2}, the deflection with a probability of 0.9973 does not exceed 8 *mm*.

In **Table 9**, k_{max} is the utilization factor of the steel beam section; $q_v{}^{2mm}$, $q_v{}^{8mm}$ – respectively, the load at which the deflection of the beam is 2 and 8 *mm*; f is the deflection of the steel beam under the design load.

In **Table 10**, $n_{свес}$ is the number of eaves included in the calculation ($n_{свес} = 1$ because the reinforced concrete deck rests on the beam on one side); $J_{tot,cb}$ is the

Beam brand	*B, m*	*l, m*	*h, mm*	*b, mm*	*s, mm*	*t, mm*	J_{tot}*, m*^{4}	*w, m*^{3}	k_{max}	$q_v{}^{2mm}$*, kN/m*^{2}	$q_v{}^{8mm}$*, kN/m*^{2}	*f, mm*
B-12	0.90	10.40	350.0	300.0	10.0	16.0	0.0003	0.0039	0.215	0.88	3.44	21.1
B-45	5.40	14.71	700.0	300.0	14.0	30.0	0.0023	0.0118	0.320	0.54	2.17	30.7
B-49	5.40	14.71	700.0	300.0	14.0	24.0	0.0020	0.0102	0.375	0.46	1.84	36.0
B-61	5.40	10.40	630.0	300.0	12.0	14.0	0.0010	0.0064	0.605	0.49	1.98	33.0

Table 9.
Geometric parameters of composite steel and reinforced concrete beams.

Beam brand	$n_{overhang}$, pieces	$J_{tot,cb}$, m^4	E_{red}, kN/m^2	k_{max}^c	q_v^{2mm}, kN/m^2	q_v^{8mm}, kN/m^2	$q_v^{3\sigma}$, kN/m^2	$f_{3\sigma}$, mm	f_q, mm
B-12	1	0.0014	1.4E+08	0.009	2.45	9.82	7.75	6.3	7.4
B-45	1	0.0079	1.6E+08	0.180	1.29	5.16	5.32	3.4	12.9
B-49	1	0.0069	1.1E+08	0.207	1.13	4.52	3.02	5.3	14.6
B-61	1	0.0041	1.1E+08	0.305	1.36	5.44	4.30	6.3	12.0

Table 10.
Bearing capacity and stiffness of composite steel and reinforced concrete beams.

Beam brand	q, kN/m	$f_{TsNIISK}$	F_{Steel}	F_{CB}	q^c, kN/m	f_{Steel}	F_{CB}
1	2	3	4	5	6	7	8
B-12	4.500	2.00	11.62	3.52	7.283	18.81	5.70
B-45	4.902	3.00	3.89	1.47	10.473	8.31	3.15
B-49	4.902	4.00	3.96	1.44	12.213	9.85	3.58
B-61	11.000	2.00	8.25	2.62	20.478	15.36	4.88

Table 11.
Comparison of theoretical design parameters with experimental values of composite steel and reinforced concrete beams.

moment of inertia of the reinforced concrete beam; E_{red} is the reduced modulus of elasticity (deformation) of the reinforced concrete beam, determined by (30); k_{max}^c is the utilization factor of the steel concrete beam section; $f_{3\sigma}$ – design deflection with security 3σ under live load $q_v^{3\sigma}$; f_q is the deflection of the reinforced concrete beam under the total design load.

Table 11 compares test results with calculations.

In **Table 11** *q* is the test load; q^c – design load; $f_{TsNIISK}$ – deflections recorded during the test; f_{Steel} and f_{CB} are design deflections of steel and reinforced concrete beams, respectively. During testing, the screed made of fine-grained concrete of class B20 with a thickness of 60 to 80 *mm* was not removed. In the calculations, the effect of the screed was not taken into account. This explains the somewhat overestimated calculated deflections in comparison with the experimental ones.

5.1 Conclusions

Taking into account the joint work of steel beams with reinforced concrete flooring allows to increase the bearing capacity and rigidity of structures by 1.4 and 3.3 times, respectively, while meeting the requirements for two groups of limit states with a given probability.

6. General conclusions

1. The use of probabilistic methods in the design of building structures, in particular, the linearization method, makes it possible to reduce their cost and shows the unequal reliability of structures designed according to current building regulations.

2. The proposed parameters for regulating the structure of the reliability of any technical system make it possible to design a building object, in particular, a three-span continuous beam with a given reliability. This will thus reduce the cost of the construction project.

3. The use of the coefficient of parallel operation of the reinforcement makes it possible to obtain a more accurate assessment of the probability of failure-free operation of the structure (the value of its reliability) and the building as a whole.

   - With excessive reliability, we save resources or increase the allowable load on the structure.

   - In case of insufficient reliability of the structure, we can take timely measures to prevent the structure from going beyond the limits of the limit state.

4. The joint work of steel beams with a reinforced concrete slab allows to increase the bearing capacity and rigidity of structures, which significantly reduces the cost of a construction project.

Author details

Andrey Dolganov
STB-Project, Moscow, Russia

*Address all correspondence to: dolganov-58@mail.ru

References

[1] Dolganov AI, Sakharov AV. On the assignment of dependability level. Dependability, 2018**18**(3). (in Russ.); DOI: 10.21683/1729-2646-2018-18-3-18-21

[2] Dolganov AI. On the security of the ice load in the Gulf of Finland. In: Materials of the 9th International Scientific and Practical Conference / Scientific Council of the Russian Academy of Sciences, Federal Agency of Scientific Organizations, EMERCOM of Russia, Russian Foundation for Basic Research. Vol. 2. Moscow: Peoples Friendship University of Russia; 2015. pp. 100-106. (in Russ)

[3] Dolganov AI, Danielov ER. Problems of optimal design of structures. Sat. reports of the III-th All-Russia. seminar: In 2 volumes. - Novosibirsk: NGASU; 2000. p. 1. (in Russ.)

[4] Dolganov AI. Optimization of reinforced concrete bridge beams by the criterion of reliability. In: Problems of optimal design of structures: Sat. reports of the II All-Russian seminar: Novosibirsk: NGASU; 1998. (in Russ.)

[5] Dolganov AI. Assessment of reliability of monolithic multi-storey buildings. Industrial and civil construction. 2010;**8**50-51. (in Russ.). DOI: 10.33622/0869-7019; ISSN 0869-7019.

[6] Probabilistic Methods in Structural Engineering (G. Augusti, A. Baratta and F. Casciati). London New York: Chapman and Hall; 1984. p. 584. ISBN 5-274-00212-9

[7] Dolganov AI. Reliability of core reinforced concrete structures. Magadan: MAOBTI; 2001. p. 209. (in Russ.)

Chapter 10

Perspective Chapter: New Theoretical Basics of Calculation of Reinforced Concrete Elements

Khanlar Seyfullaev

Abstract

In this chapter, on the basis of the nonlinear deformation model of the mechanics of deformable solids, new theoretical foundations for the design of reinforced concrete elements have been studied. This is the first formulation of the strength conditions of reinforced concrete elements relative to the height of the compressed concrete zone ξ in the following form: for eccentrically tensioned elements $0 \le \xi < \xi_{OR}$, where $\xi_{OR} = 1/\left(1 + \frac{\varepsilon_{sult}}{\varepsilon_{bult}}\right)$, for bending elements $\xi_{OR} \le \xi \le \xi_R$, where $\xi_R = 1/\left(1 + \frac{\varepsilon_{sl}}{\varepsilon_{bult}}\right)$ for eccentrically compressed elements with three cases of calculation: (1) $\xi_{OR} \le \xi \le \xi_R$; (2) $\xi_R < \xi < 1$; and (3) $1 \le \xi \le \infty$. The height of the compressed zone of concrete ξ is determined from the condition $M_X \le M_{X,ult}$. The representation of strength conditions in the above form makes it possible to develop new and accurate solutions to the elastic-creep theory of mechanics of reinforced concrete, taking into account the real properties of reinforced concrete materials. The research concluded several important points; for example, it has been established that the developed method for designing reinforced concrete elements is based on an accurate and rigorous application of the nonlinear deformation model of the mechanics of deformed solids to the problems of elastoplastic bending of the theory of reinforced concrete. The developed method for designing reinforced concrete elements makes it possible to take into account the main characteristics of materials: the long-term strength of concrete and the increase in the strength of reinforcement (rebar) after the yield point in accordance with the state diagrams of concrete and reinforcement compared to other important matters worth publishing to shed light on.

Keywords: Eurocode-2, strength conditions, height of the compressed zone of concrete, nonlinear deformation model, plastic hinge model, new calculation method

1. Introduction

The rules and principles established by the Eurocodes are a strict set of design provisions on the basis of which the design of reinforced concrete structures should be carried out. The European reinforced concrete standards have incorporated numerous scientific research and experience of prominent scientists from various countries, including Russia, the best sections of the national standards of these countries, a reasoned formulation of the main hypotheses, and generally accepted methods.

After the approval of this set as the standards of the European Community, the theory and practice of reinforced concrete begin to develop within the framework of strict compliance with all the requirements, design provisions of the Eurocode deformation model of the section.

It should be noted that the inconsistency between the method of limit states of the Eurocode and the method of limit loads of the Russian standard was repeatedly pointed out by the creators of the standard K.E. Tal and A.A. Gvozdev [1]. Comparing the national standard, now the basis of the updated SNiP 52.01-2003, and European standards, A.A. Gvozdev pointed out their significant difference in the principles and methods of calculation regarding the design of normal sections and the action of duration of the load on reinforced concrete structures.

In the methods and limit states of the national standard, the hypothesis of flat sections, one of the main hypotheses in the Eurocode and the classical theory of mechanics, is categorically rejected.

To determine the creep deformation of concrete, there are two theories based on the basic properties of concrete deformation: the theory of aging and the theory of heredity.

As experiments show, the theory of aging exaggerates the influence of aging of materials and underestimates the calculated values of creep deformations.

Studies to test the theory of heredity led to the conclusion that it not only excludes aging, but also exaggerates the calculated values of creep deformations. As noted by V. M. Bondarenko [2], both of these theories give some limit values of creep deformations: the smallest and the largest. In addition, as the analysis showed, in some cases their use can lead to qualitatively incorrect results. The main reason for their existence and application is the relative mathematical simplicity, which consists in an elementary transition from integral to differential equations of state [2]. To eliminate the shortcomings of mentioned theories of creep, the Soviet scientist N.G. Maslov [2] proposed the theory of elastic-plastic (creeping) body of mechanics of solid deformed bodies, the essence of which is that this theory takes into account both the properties of concrete aging and the properties of heredity.

Considering the above, in this chapter, the state diagrams of concrete and reinforcement, taking into account elastic-plastic deformations, are summarized and presented in the form of a two-line diagram (**Figure 1**).

Thus, the study of elastic-creep bending of reinforced concrete elements is reduced to solving the generally accepted method of the theory of plasticity (creep) of deformed solid bodies of mechanics.

For a compressed concrete prism in the mode of proportional development of longitudinal deformations in time, a gradual decrease in the resistance of concrete is detected, the so-called descending branch of the stress-strain diagram (**Figure 1c**).

The greatest intensity of growth of inelastic deformations is observed in the first 3-4 months and may continue for several years.

In the diagram, section O-A characterizes the deformations that occur during loading, and section A-B characterizes the rising of inelastic deformations (creep) at a constant value of stresses.

According to experimental data, under a long-term load under the influence of developing significant inelastic deformations and structural changes, concrete is

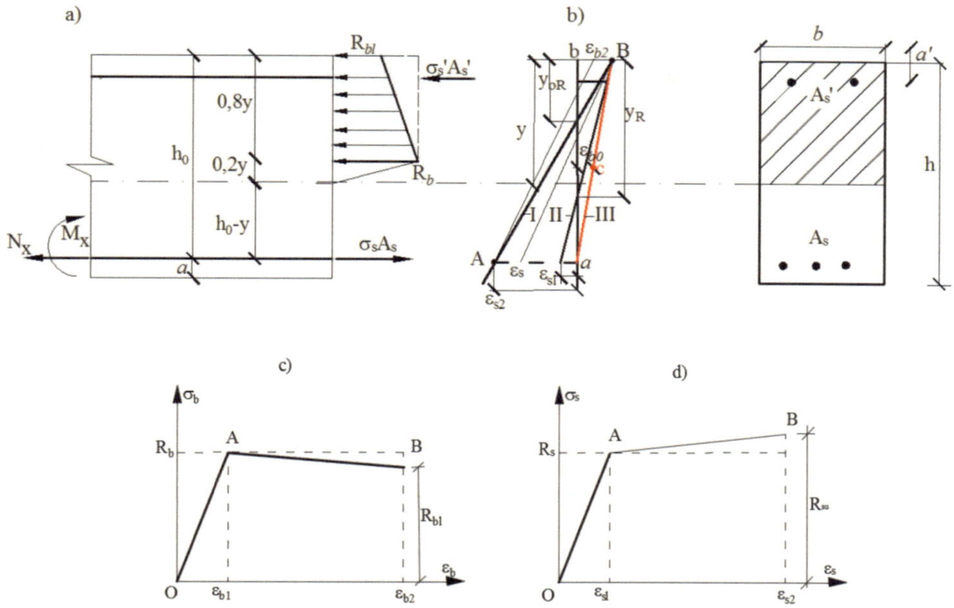

Figure 1.
Design scheme of bending elements, (a) diagram of normal stresses taking into account creep and long-term strength of the compressed zone of concrete, (b) deformation diagrams of a flat section at various limit states: I: at eccentric tension; II: at bending; III: at eccentric compression, (c) state diagram of concrete taking into account creep and long-term strength of concrete, and (d) diagram of the state of rebar, taking into account the increase in the strength of rebar after the yield point.

destroyed at stresses less than the temporary resistance to axial compression R_b; $R_{bl} = 0,9R_b$ and less than [3].

In 2014, the Russian newspaper "Stroitelnaya Gazeta" published an article "Inconsistency between the updated standard for reinforced concrete and the Eurocode - an obstacle in construction" [1]. Because the national standard for reinforced concrete AzDTN 2.16-1 was developed on the basis of the updated standard for reinforced concrete SNiP 52.01-2003 of Russia, this issue was studied at AzNIISA and a decision was made to bring it to conformity with Eurocodes. These differences cannot be eliminated by any correction or conversion coefficient.

When comparing national and European standards, their significant differences in the concepts of limit states and methods for designing reinforced concrete elements were revealed.

Therefore, instead of the limit state, which is proceeded from the stage of destruction of the stressed state of bending elements, a new form of limit states is proposed in the form of strain diagrams obtained on the basis of the hypothesis of flat sections. The essence and form of these limit states are described and given in Refs. [2–7].

Limit states of bending elements in accordance with the nonlinear deformation model are accepted in the form of strain diagrams of flat sections, which pass through one of these three characteristic points of the section, corresponding to deformations of tensed reinforcement $\varepsilon_{s,max} = \varepsilon_{s,2}$ under eccentric tension, ultimate deformation of the compressed zone of concrete $\varepsilon_{b,max} = \varepsilon_{b,2}$ under simple bending, as well as $\varepsilon_b = \varepsilon_{b0}$ under eccentric compression [4, 8, 9].

As is known, the only correct solution to the problem of mechanics of deformable solids will be when all three aspects of the problem: static, geometric, and physical are combined into a single one. In this formulation, on the basis of the nonlinear deformation model, the problem of elastoplastic bending of reinforced concrete elements was solved and an analytical solution of this problem is given in Refs. [3, 7].

The limit states for the strength of bendable reinforced concrete elements, adopted in Refs. [3, 4], based on a nonlinear deformation model of the mechanics of deformable solids, make it possible to compile strength conditions in a general form, taking into account the main deformation characteristics: both creep and long-term strength of concrete and the increase in the strength of reinforcement after the yield point, the state diagrams of which are shown in **Figure 1c** and **d**.

Two-line diagram of concrete:

at $0 > \varepsilon_b \le \varepsilon_{b1}$ $\sigma_b = E_b \cdot \varepsilon_b$;

at $\varepsilon_{b1} \le \varepsilon_b \le \varepsilon_{b2}$ $\sigma_b = R_b\left[\left(1 - \frac{\varepsilon_b - \varepsilon_{b1}}{\varepsilon_{b2} - \varepsilon_{b1}}\right) + \frac{R_{bl}}{R_b}\frac{\varepsilon_b - \varepsilon_{b1}}{\varepsilon_{b2} - \varepsilon_{b1}}\right] \le R_b$.

Two-line diagram of rebar:

at $0 > \varepsilon_s \le \varepsilon_{sl}$ $\sigma_s = \varepsilon_s \cdot E_s$;.

at $\varepsilon_{sl} \le \varepsilon_s \le \varepsilon_{s2}$ $\sigma_s = R_s\left[\left(1 - \frac{\varepsilon_s - \varepsilon_{sl}}{\varepsilon_{s2} - \varepsilon_{sl}}\right) + \frac{R_{su}}{R_s}\frac{\varepsilon_s - \varepsilon_{sl}}{\varepsilon_{s2} - \varepsilon_{sl}}\right] \ge R_s$.

The strength conditions of reinforced concrete elements are as follows:

$$\varepsilon_{b,\max} \le \varepsilon_{b,ult}; \varepsilon_{s,\max} \le \varepsilon_{s,ult} \quad (1)$$

where $\varepsilon_{b,ult} = \varepsilon_{b2}$ is the ultimate relative deformation of concrete in compression: $\varepsilon_{s,ult} = \varepsilon_{s2}$ ultimate relative deformation of reinforcement at breaking.

Depending on the new forms of limit states, a method for designing reinforced concrete bending elements has been developed that meets the main requirements of Eurocodes.

2. The novelty of this work

The novelty of this work lies in the fact that the developed method for calculating reinforced concrete elements for the first time takes into account the long-term strength of concrete and the increase in the strength of reinforcement after the yield point, the account of which leads to savings in reinforcement.

The purpose of developing a methodology for calculating reinforced concrete elements is that its application in existing regulatory documents will ensure their complete compatibility with Eurocodes.

Then, the solution of the problem of bending elements based on a nonlinear deformation model with piecewise linear state diagrams of concrete is reduced to solving the following static equations, subject to the hypothesis of flat sections and analytical expressions of a two-line concrete state diagram (**Figure 1c**):

1.$M = \int_{A_b}\sigma_b bydy + \int_{A_s}\sigma_s(h_0 - y)dA_s + \int_{A_s'}\sigma_s'(y - a')dA_s'$;

$$\int_{A_b}\sigma_b bdy - \int_{A_s}\sigma_s dA_s + \int_{A_s'}\sigma_s' dA_s' = 0;$$

where

$$\int_{A_b}\sigma_b bydy = \int_0^{y_0}\varepsilon_b \bullet E_b bydy + \int_{y_0}^{y} R_b(\varepsilon_b)bydy$$

$$\int_{A_b} \sigma_b b dy = \int_0^{y_0} \varepsilon_b \bullet E_b b dy + \int_{y_0}^{y} R_b(\varepsilon_b) b dy$$

2. Linear deformation diagram obtained on the basis of the hypothesis of flat sections, where, at a known value of the ultimate deformation of concrete ε_{b2}, the deformations of the reinforcement ε_s and ε_s' are determined, and then, the stresses at the characteristic points of the concrete's compressed zone section;

3. Piecewise linear or curvilinear state diagrams of concrete characterizing the deformations and stresses of the compressed zone of concrete.

As a result, all aspects of the problem are combined: statistical, geometrical (deformational), and physical.

The application of a mathematically rigorous nonlinear deformation model of the mechanics of deformable solids to the problems of the theory of reinforced concrete made it possible to propose a new version of the limit states of bending elements and, on their basis, develop calculation methods for designing reinforced concrete structures, which is one of the pressing problems of structural mechanics.

3. The content of the work

Due to the fact that the strength conditions (1) are general, depending on the purpose and problem setting, they can be presented in a more convenient form, for example, within the limits of elasticity of concrete and reinforcement, based on Hooke's law, they are easily expressed in the form of the corresponding well-known theory of allowable stresses:

$$\sigma_{b,\max} \le R_b; \sigma_{s,\max} \le R_s \tag{2}$$

When applying the theory of the nonlinear deformation model of the mechanics of deformable solid bodies to reinforced concrete structures, based on the following formulas of resistance of materials [10–13]:
$\frac{1}{r} = \frac{\varepsilon_{b,\max}}{y} = \frac{M_x}{D}$ we find:
$\varepsilon_{b,\max} = \frac{M_x y}{D}$ and $\varepsilon_{b,ult} = \frac{M_{x,ult} y}{D}$, instead of (1) we get:

$$M_x \le M_{x,ult}$$

On the other hand, depending on the limit states for different types of deformations, the strength conditions (1) are simplified and take a simpler form:

According to the limit state for eccentrically tensed elements, the limit state is accepted in the form where the relative elongation of the reinforcement is taken to be equal $\varepsilon_{s,ult}$, the strength condition is satisfied $\varepsilon_{s,\max} \le \varepsilon_{sult}$, based on this condition, and the concrete deformation is found as follows:

$$\varepsilon_b = \varepsilon_{s,ult} \frac{y}{h_0 - y} \le \varepsilon_{b,ult}$$

we find from here: $y \leq y_{OR}$ or $\xi \leq \xi_{OR}$
where $\xi_{OR} = \frac{1}{1+\frac{\varepsilon_{s,ult}}{\varepsilon_{b,ult}}}$
Thus, for eccentrically tensed elements, strength conditions (1) are satisfied if:

$$0 \leq \xi \leq \xi_{OR} \quad (3)$$

Analogously, according to the limit state for simple bending, where the largest relative deformation of the most compressed concrete fibers is taken to be equal $\varepsilon_{b,ult}$, then the condition $\varepsilon_{b,\max} \leq \varepsilon_{b,ult}$ is satisfied identically, and based on this condition, the relative deformation of the stretched reinforcement is found as:
$\varepsilon_s = \varepsilon_{b,ult} \frac{h_0 - y}{y} \leq \varepsilon_{s,ult}$, and from here we find:

$$y \geq y_{OR} \text{ or } \xi \geq \xi_{OR}$$

On the other hand, at bending, another condition is fulfilled $\xi \leq \xi_R$, where $\xi_R = \frac{1}{1+\frac{\varepsilon_{sl}}{\varepsilon_{b,ult}}}$, then we have:

$$\xi_{OR} \leq \xi \leq \xi_R \quad (4)$$

There are three cases for eccentrically compressed elements, and the strength conditions for each case take the following form:

a. $\xi_{OR} \leq \xi \leq \xi_R$, as for bending elements;

b.

$$\xi_R \leq \xi \leq 1; \quad (5)$$

c. $1 \leq \xi \leq \infty$, for which the calculation method will be developed separately in a special form.

In all considered cases, the strength check is reduced to compliance with the condition (3)–(5) where it is required to determine the height of the compressed zone of concrete, which is determined from the following condition:

$$M_x \leq M_{x,ult} \quad (6)$$

When comparing the strength conditions compiled on the base of the nonlinear deformation model in the form (3)–(5) with the method based on the plastic hinge model, we see that we have only one strength condition:

$$M_x \leq M_{x,ult}$$

And the second strength condition (1) for tensed reinforcement $\varepsilon_{s,\max} \leq \varepsilon_{sult}$ cannot be established because the law of flat sections is rejected, which leads to inconsistencies with Eurocodes.

Verification of the fulfillment of the strength conditions (3)–(5) is related to the height of the compressed zone of the concrete of bending elements, which is determined from the condition (6), where $M_{x,ult}$ has the form [3]:

$$M_x \le M_{x,ult} =$$
$$= R_b b h_0^2 \left\{ \begin{array}{l} (1-k_0)\xi\left[1-\frac{1}{2}(1-k_0)\xi\right]+\frac{1}{2}k_0\xi\left[1-\left(1-\frac{2}{3}k_0\right)\xi\right]- \\ -\frac{1}{2}(1-\gamma_{bl})(1-k_0)\xi\left[1-\frac{1}{3}(1-k_0)\xi\right] \end{array} \right\}+$$
$$+\sigma_s' A_s'(h_0-a'); \quad (7)$$
$$N=\sigma_s A_s-\sigma_s' A'-R_b b h_0\xi\left[\left(1-\frac{k_0}{2}\right)-\frac{1}{2}(1-k_0)(1-\gamma_{bl})\right]$$

here $M_{x,ult}$—limiting moment of internal forces relative to the center of gravity of tensioned reinforcement, the value of which is determined in Ref. [3];

$\xi = \frac{y}{h_0}$—height of the compressed zone of concrete;

$\gamma_{bl} = \frac{R_{bl}}{R_b}$, R_{bl}—long-term strength of concrete;

$k_0 = \frac{\varepsilon_{b,1}}{\varepsilon_{b,\max}}$—coefficient of elastoplastic deformation of concrete, which varies within: $k_0 = 0,15 - 1,0;$ and the long-term strength of concrete $R_{bl} = \gamma_{bl}R_b$ и $\gamma_{bl} = 0,8\div1,0$. By varying these coefficients, we find the values of the limiting moment $M_{x,ult}$ and the shapes of the stress diagram in concrete in accordance with $M_{x,ult}$ (**Figure 1a**).

$k_0 = 1,$ in this case, concrete and reinforcement work in the elastic stage, therefore, $\gamma_{bl} = 1$ and concrete creep does not occur.

$A_R = 0,5\xi_R\left(1-\frac{1}{3}\xi_R\right) = 0,208$ and therefore $M_{xult} = 0,208R_b b h_0^2$, which differs from the solution in the limit state by 46.8%.

$k_0 = 0,15 - 1,0,$ limiting moment reaches its maximum value at $k_0 = 0,2$ and minimum at $\gamma_{bl} = 0,85$:

$$M_{x,ult} = R_b b h_0^2 A_R$$

where

$$A_R = 0,84\xi_R - 0,3907\xi_R^2 = 0,3865$$

If we neglect the elastic deformation of concrete near the neutral layer of the section, then we have:

$A_R = 0,74\ \xi_R - 0,304\ \xi_R^2 = 0,3583$ and the difference is about 6.7%; therefore, under conditions (7), we neglect the influence of elastic deformations of concrete in the compressed zone of the element and have:

$A_R = 0,8\ \xi_R - 0,32\ \xi_R^2 = 0,3912$ and the difference, taking into account long-term strength, is 9.2%.

Thus, the design scheme of bending elements is accepted as shown in **Figure 1a**, where the influence of the long-term strength of concrete is taken into account, which is fully consistent with the principles and requirements of the Eurocode.

Based on the accepted design scheme, the strength conditions (6) take the following form:

$$M \le R_b b h_0^2 A_0 + \sigma_s' A_s'(h_0 - a') \quad (8)$$
$$N = \sigma_s A_s - \sigma_s' A_s' - 0,8R_b b h_0\xi[1-0,4(1-\gamma_{bl})]$$

where

$$A_0 = 0,8\xi(1-0,4\xi) - 0,4(1-\gamma_{bl})\xi(1-0,267\xi)$$

Strength conditions (8) are valid if conditions (3)-(5) are observed, written for various types of bending of reinforced concrete elements.

A new view at the limit states and strength conditions (1) made it possible to develop methods for calculating reinforced concrete elements for various forms of bending.

Having determined $A_0 = \frac{M}{R_b b h_0{}^2}$, from the condition (8), for determining the height of the compressed zone we obtain the following equation:

$$\xi^2 - \frac{0,8-0,4(1-\gamma_{bl})}{0,32-0,1068(1-\gamma_{bl})}\xi + \\ + \frac{M}{R_b b h_0{}^2[0,32-0,1068(1-\gamma_{bl})]} = 0 \quad (9)$$

After finding ξ, the conditions (3)–(5) written for each type of bending of reinforced concrete elements are checked. The following technique is proposed for finding the stresses in the reinforcements, taking into account the increase in stresses after the yield point.

4. Design according to the first group of limit states

4.1 Eccentric tension

In accordance with the limit state, for this type of deformation in tensed reinforcement, the relative deformation is equal to $\varepsilon_{s,ult}$, therefore, the stress $\sigma_s = R_{su}$, and in the compressed zone $\varepsilon_s' = \varepsilon_{s,ult}\frac{y-a'}{h_0-y}$, and the stress in accordance with ε_s' is determined as follows (**Figure 2**):

at $\varepsilon_s' \leq \varepsilon_{sl}$, $\sigma_s' = E_s\varepsilon_s$ and at $\varepsilon_s' > \varepsilon_{sl}$

$$\sigma_s' = R_s\left[\left(1-\frac{\varepsilon_s'-\varepsilon_{sl}}{\varepsilon_{s2}-\varepsilon_{sl}}\right) + \frac{R_{su}}{R_s}\frac{\varepsilon_s'-\varepsilon_{sl}}{\varepsilon_{s2}-\varepsilon_{sl}}\right]$$

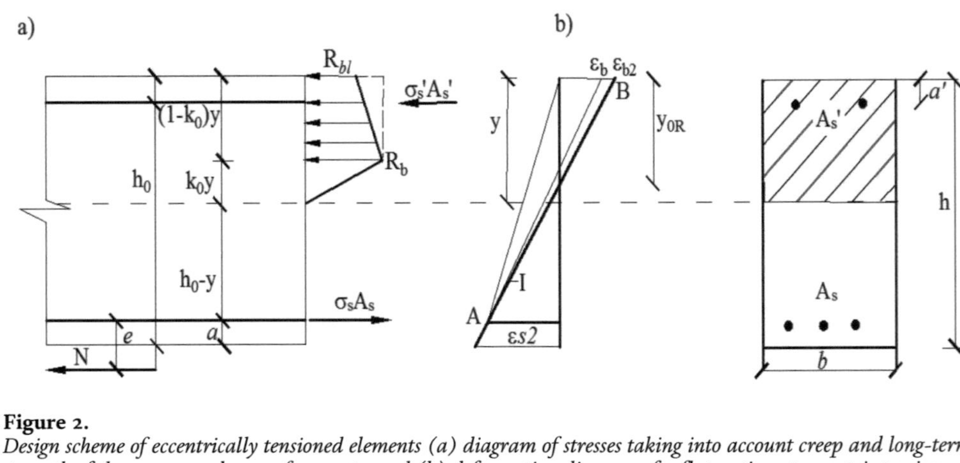

Figure 2.
Design scheme of eccentrically tensioned elements (a) diagram of stresses taking into account creep and long-term strength of the compressed zone of concrete, and (b) deformation diagram of a flat section at eccentric tension.

The condition $\xi \leq \xi_{OR}$is checked. If this condition is satisfied, then $A'_s = 0$ and the reinforcement area in the tension zone is determined as follows.

$$A_s = \frac{1}{R_{su}}\left\{\frac{N}{100} + 0,8\xi R_b bh_0[1 - 0,4(1 - \gamma_{bl})]\right\}$$

If $\xi > \xi_{OR}$, then the destruction of the element will be along the compressed zone of concrete. In order to prevent destruction, it is necessary to ensure $\xi \leq \xi_{OR}$.

Assuming $\xi = \xi_{OR}$ we find the minimum value ofA'_s:

$$A'_s = \frac{N \cdot e - A_{OR} R_b bh_0{}^2}{\sigma'_s(h_0 - a')} \text{ and } A_s = \frac{1}{R_{su}}\left\{\frac{N}{100} + 0,8\xi_{OR} R_b bh_0[1 - 0,4(1 - \gamma_{bl})]\right\} + A'_s\frac{\sigma'_s}{\sigma'_s}$$

4.2 Simple bending of reinforced concrete elements

At simple bending, the conditions $\xi_{OR} \leq \xi \leq \xi_R$must be met. The value of ξ is determined by solving the quadratic Eq. (7).

In contrast to eccentrically tensed elements, here, according to the limit state, the relative deformation of the most compressed extreme fibers of concrete is taken equal to $\varepsilon_{b,ult}$ and the condition $\varepsilon_{b,\max} \leq \varepsilon_{b,ult}$ is satisfied identically.

Then, the deformations of the reinforcement are determined as follows:

$$\varepsilon_s = \varepsilon_{b,ult}\frac{1-\xi}{\xi};\ \varepsilon'_s = \varepsilon_{b,ult}\frac{\xi - \delta'}{\xi};\ \delta' = \frac{a'}{h_0}$$

Further, depending on the value of ε_s and ε'_s, the stresses in the reinforcements A_s and $A'_s$are determined:

If $\varepsilon_s \geq \varepsilon_{sl}$ and $\varepsilon'_s > \varepsilon_{sl}$, then the stresses in the reinforcements will be $\sigma_s = R_s\left[\left(1 - \frac{\varepsilon_s - \varepsilon_{sl}}{\varepsilon_{s2} - \varepsilon_{sl}}\right) + \frac{R_{su}}{R_s}\frac{\varepsilon_s - \varepsilon_{sl}}{\varepsilon_{s2} - \varepsilon_{sl}}\right]$ and by the same formula σ'_s at ε'_s.

When determining areas $A_s$and A'_s there may be the following cases:

a. the conditions $\xi \geq \xi_{OR}$ and $\xi \leq \xi_R$, are met and then reinforcement is not required in the compressed zone of concrete $A'_s = 0$; .

In the stretched zone A_s is determined by the following formula:

$$A_s = \frac{1}{\sigma_s} 0,8R_b bh_0\xi[1 - 0,4(1 - \gamma_{bl})]$$

If the condition $\xi \geq \xi_{OR}$ is not met, then choosing A'_s structurally or assuming $\xi = \xi_{OR}$ by design it is ensured the equation $\xi = \xi_{OR}$ and the areas of reinforcements are determined as follows:

$$A'_s = \frac{M - A_{OR} R_b bh_0{}^2}{\sigma'_s(h_0 - a')};$$

$$A_s = \frac{1}{\sigma_s} 0,8R_b bh_0\xi[1 - 0,4(1 - \gamma_{bl})] + A'_{so}\frac{\sigma'_s}{\sigma_s}$$

b. When $\xi > \xi_R$, is observed, the condition $\xi > \xi_{OR}$is identically satisfied. The reinforcement area is found as a double reinforcement:

$$A_s' = \frac{M - A_R R_b b h_0{}^2}{\sigma_s'(h_0 - a')}$$

$$A_s = \frac{1}{\sigma_s} 0,8 R_b b h_0 \xi [1 - 0,4(1 - \gamma_{bl})] + A_s' \frac{\sigma_s'}{\sigma_s}$$

4.3 Eccentric compression

As noted above, there are three cases in the design of eccentric compression (5).

The strength conditions for the first two cases of changing the height of the compressed zone have the form (8). Under these strength conditions, there are four unknown parameters ξ, σ_s, A_s' и A_s, a $\sigma_s'$and are determined as follows:

$\varepsilon_s' = \varepsilon_{b,ult} \frac{\xi - \delta'}{\xi}$, where $\varepsilon_s' > \varepsilon_{sl}$. And stresses:

$$\sigma_s' = R_s \left[\left(1 - \frac{\varepsilon_s' - \varepsilon_{sl}}{\varepsilon_{s2} - \varepsilon_{sl}} \right) + \frac{\varepsilon_s' - \varepsilon_{sl}}{\varepsilon_{s2} - \varepsilon_{sl}} \cdot \frac{R_{su}}{R_s} \right]$$

e—the eccentricity of the normal force and for a rectangular section is equal to:

$$e = e_0 \eta + \frac{h_0 - a'}{2},$$

η—coefficient taking into account the longitudinal bending:

$$\eta = \frac{1}{1 - \frac{N}{N_{cr}}}$$

N_{cr}—Euler's critical force and it is equal to:

$$N_{cr} = \frac{\pi^2}{l_0^2} D \qquad (10)$$

where *D*—bending stiffness is defined as follows [11]:

$$D = \frac{M_x y}{\varepsilon_{b,\max}}$$

Here, M_x is determined by the formula (7).

D—the stiffness of the element in bending, taking into account the long-term strength of concrete and the stressed state, the value of which was determined in Ref. [3]:

$$D = k_b E_b J_{x,b} + k_s E_s J_{x,s}; \qquad (11)$$

here: $k_b = 1,5 k_0 (1 - \frac{1}{3} k_0{}^2) - (1 - \gamma_{bl})(1 - k_0)(1 + \frac{k_0}{2}) k_0$,

$$k_s = \frac{\varepsilon_{sl}}{\varepsilon_s k_0}; J_{x,b} = \frac{b y^3}{3}; J_{x,s} = n A_s (h_0 - y)^2 + n A_s' \frac{\varepsilon_s'}{\varepsilon_s} (y - a')^2$$

$\varepsilon_s = \varepsilon_{b2}\frac{1-\xi}{\xi}, \varepsilon_s' = \varepsilon_{b2}\frac{\xi-\delta'}{\xi}$

In view of the fact that two strength conditions (8) contain four unknowns, a special calculation technique is required to solve the problem.

In order to reduce the number of unknowns, the following form of replacement of unknowns is proposed [4].

$$N + \sigma_s A_s = R_s A; A_s' = A'; M = N \cdot e \quad (12)$$

After replacement, the strength conditions (8) take the following form:

$$\begin{gathered} M = R_b b h_0{}^2 A_0 + \sigma_s' A_s'(h_0 - a'); \\ R_s A = R_b b h_0 \xi[0,8 - 0,4(1-\gamma_{bl})] + \sigma_s' A' \end{gathered} \quad (13)$$

This system completely coincides with the strength equations of elements in pure bending with double reinforcement. Therefore, assuming in them $\xi = \xi_R$, we find the cross-sectional area A and A'.

$$A' = \frac{M - R_b b h_0{}^2 A_R}{\overline{\sigma'}_s(h_0 - a')};$$

$$A = \frac{1}{R_s} R_b b h_0 \xi_R[0,8 - 0,4(1-\gamma_{bl})] + \frac{\sigma_s'}{R_s} A';$$

Using conditions (12), we find A and A_s'.

$$A_s' = A' = \frac{M - R_b b h_0{}^2 A_R}{\overline{\sigma'}_s(h_0 - a')}; A_s = \frac{R_s}{\sigma_s}\left(A - \frac{N}{100R_s}\right)$$

This technique is also valid for the first case of calculation, when $\xi_{OR} \le \xi \le \xi_R$ here at $A - \frac{N}{100R_s} \le 0, A_s$ has a negative value and $A_s = 0$. Then we have $A_s = 0; A_s' \ne 0$.

Theoretically, it could be $A_s = 0; A_s' = 0$. This is possible with $N \cdot e \le R_b b h_0{}^2 A_R$ and $A \le \frac{N}{100R_s}$. The section should be reinforced structurally.

When $A_s = 0$, it is required to determine the height of the compressed concrete zone. To this end, from the second equation we find:

$$\sigma_s' A' = R_s A - R_b b h_0 \xi[0,8 - 0,4(1-\gamma_{bl})]$$

and then substituting it into the first equation, we get.

$$\begin{gathered} M = R_b b h_2^0 A_0 - \\ -\{R_s A - R_b b h_0 \xi[0,8 - 0,4(1-\gamma_{bl})]\}(h_0 - a') \end{gathered}$$

From this equation for finding ξ we have:

$$\begin{gathered} \xi^2 - \frac{a'}{h_0}\frac{0,8 - 0,4(1-\gamma_{bl})}{0,32 - 0,1068(1-\gamma_{bl})}\xi - \\ -\frac{N}{R_b b h_0{}^2[0,32 - 0,1068(1-\gamma_{bl})]} = 0 \end{gathered} \quad (14)$$

This solution is valid for the second case of calculation (5), when $\xi_R \leq \xi \leq 1$.

Having solved the equations, we find ξ and obtain that $\xi > \xi_R$, at which the reinforcement area $A_s = 0$, we find:

$$A_s = 0, A_s' = \frac{N \cdot e - R_b b h_0{}^2 A_0}{\sigma_s'(h_0 - a')}$$

σ_s' is determined by the method described above. Comparing the case $\xi_{OR} \leq \xi \leq \xi_R$, which is also valid for simple bending, we can conclude that in this interval at simple bending, a single reinforcement is possible, i.e., $A_s' = 0$ while at eccentric compression $A_s' > A_s$ and we always have $A_s' \neq 0$, $A_s \neq 0$, for which $\xi = \xi_R$.

At eccentric compression, the third case of calculation is possible, when $1 \leq \xi < \infty$. In this case, another calculation method is required that corresponds to the strain diagram for new limit states.

When the section is fully compressed, in accordance with the limit state, the deformation diagram passes through the point "C" where the concrete deformation ε_b is equal to $\varepsilon_{b0} = 0,002$, and the stress diagram in concrete has the form of a trapezoid-parabola (**Figure 3**) [4].

It should be noted that we solved this problem without taking into account the long-term strength of concrete $\gamma_{bl} = 1$, in work [4], only the final results of the calculation are given here.

$$S = \psi R_b bh; \psi = \overline{\gamma}_{bl} - \frac{3,0416}{\left(\frac{7y}{h} - 3\right)^2}; \delta = \frac{6}{7} - \frac{1}{49\psi}(10 + 7,5\gamma_{bl}); \overline{\gamma}_{bl} = \frac{1}{7}[4 + 1,5(1 + \gamma_{bl})]$$

When "y" varies within $h \leq y < \infty$, value of coefficient ψ varies within $(\overline{\gamma}_{bl} - 0,19) \leq \psi \leq \overline{\gamma}_{bl}$, and at $\gamma_{bl} = 0,85$; $0,778 \leq \psi \leq 0,968$

The strength conditions drawn up relative to the neutral axis of the section take the form:

$$M \leq \psi R_b bh(y - \delta h) + \sigma_s' A_s'(y - a') \qquad (15)$$
$$N = \psi R_b bh + \sigma_s A_s + \sigma_s' A_s'$$

The stress values in the reinforcements are determined by the values of ε_s и ε_s' which were found from the strain diagram depending on ε_{b0}.

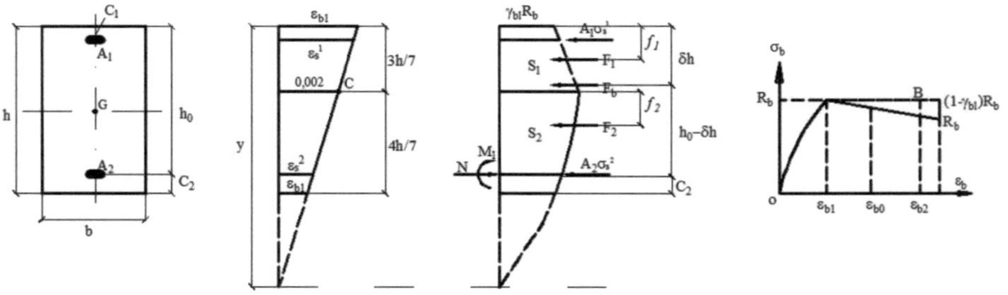

Figure 3.
Design scheme of eccentrically compressed elements (a) deformation diagram of a flat section, (b) diagram of stresses taking into account long-term strength of the compressed zone of concrete, and (c) state diagram of concrete taking into account creep and long-term strength of concrete.

When $y = \infty$, all the fibers of the section experience the same deformations $\varepsilon_{b0} = 0,002$ and the stresses will be equal to:

$$\sigma_s = \sigma_s' = \sigma_s = R_s\left[\left(1 - \frac{\varepsilon_{b0} - \varepsilon_{sl}}{\varepsilon_{s2} - \varepsilon_{sl}}\right) + \frac{R_{su}}{R_s}\frac{\varepsilon_{b0} - \varepsilon_{sl}}{\varepsilon_{s2} - \varepsilon_{sl}}\right]$$

In this case, the strength conditions will take the form:

$$M - \overline{\gamma}_{bl}(h_0 - 0,5h)R_b bh - A_s'\sigma_s'(h_0 - a') = 0 \quad (16)$$

$$N = \overline{\gamma}_{bl}R_b bh + (A_s + A_s')\sigma_s'$$

At $y = \infty$;

$$A_s' = \frac{M - (h_0 - 0,5h)\overline{\gamma}_{bl}R_b bh}{\overline{\sigma'}_s(h_0 - a')}; A_s = \frac{N - \overline{\gamma}_{bl}R_b bh}{\overline{\sigma}_s} - A_s'$$

From the condition of minimum of the sum $A_s' + A_s$is found that at $A_s' > A_s$ $A_s = 0$, and the value of ψ has the form:

$$\psi = \frac{1}{0,8571 - \frac{a'}{h}}\left[0,204(1 + 0,75\gamma_{bl}) + \frac{N(h_0 - a') - M}{R_b bh^2}\right] \quad (17)$$

and the rest of the parameters:

$$y = \frac{h}{7}\left(3 + \frac{1,7457}{\sqrt{\overline{\gamma}_{bl} - \psi}}\right); \quad (18)$$

$$1000\varepsilon_s' = 2 + \left(3,437 - 8,019\frac{a'}{h}\right)\sqrt{\overline{\gamma}_{bl} - \psi};$$

$$\varepsilon_s' = \frac{y - a'}{y - \frac{3}{7}h} \cdot \varepsilon_{b0}; \quad (19)$$

Depending on the ε_s' is found σ_s', and then $A_s = 0$;

$$A_s' = \frac{N - \psi R_b bh}{\overline{\sigma'}_s}$$

5. Calculation according to the second group of limit states

Reinforced concrete structures must also meet the requirements of the calculation for the second group of limit states, that is, suitability for normal use.

The second group of limit states according to BAEL-85 is presented in the form of the following phenomena, for which calculations are carried out:

crack formation design in the compressed zone of concrete, caused in the direction of normal stresses $\sigma_{b,max}$;

- design for the opening of cracks in the tensed zone of concrete caused by tensile stress in concrete;

- deformation design.

The design for the second group of limit states is based on the well-known method of allowable stresses, which makes it possible to determine the stresses in reinforced concrete elements using the formulas of resistance of materials:

a. The design for the formation of cracks in the compressed zone of concrete in the direction of normal stress $\sigma_{b,max}$ is carried out according to the following formula:

$$\sigma_{b,max} \leq 0,6R$$

Where R-cubic strength of concrete at axial compression is equal to the concrete class B at compression in MPa.

$\sigma_{b,max}$—the maximum stress in the compressed zone of concrete and is determined by the following formula of resistance of materials:

$$\sigma_{b,max} = \frac{M_x \bullet y_1}{J_{x,red}}$$

b. Design for the opening of cracks in the tensed zone of concrete, depending on the category of crack resistance and η coefficient of adhesion of reinforcement with concrete in reinforced concrete elements:

- opening of cracks are not harmful;

  In this case, the value of the ultimate stress in tensed reinforcement is not limited and is taken equal to $\overline{\sigma_s} = R_s$.

- opening of cracks are dangerous.

  In this case, the values of ultimate stresses in tensed reinforcement are limited by taking them equal to one of the lower values of the following condition:

$$\overline{\sigma_s} = \left(\frac{2}{3}R_s \text{ or } 150\eta\right) MPa$$

- crack opening is very dangerous.

  In this case, the values of ultimate stresses in tensed reinforcement are limited by taking them equal to one of the lower values of the following condition:

$$\overline{\sigma_s} = \left(\frac{1}{2}R_s \text{ or } 150\eta\right) MPa$$

The conditions of crack resistance of reinforced concrete elements for each of the above categories are ensured by choosing the ultimate stress $\overline{\sigma_s}$ in the design formulas of the second group of limit states when determining the area of tensed reinforcement, for example, for tensed reinforced concrete elements:

$$A_s = \frac{N_x}{\overline{\sigma_s}}$$

for bending elements with single reinforcement:

$$A_s = \frac{M_x}{\beta b \overline{\sigma_s}}$$

etc. depending on the category of crack resistance of reinforced concrete elements.

Based on the aforesaid, the values of ultimate stresses in tensed reinforcement depending on the categories of crack resistance and the type of reinforcement surface η are given in **Table 1**.

c. Deformation design:

The condition for checking is:

$$f \leq f_{ult}$$

here f—is defined as follows:

$$f = f_{cd} + f_{ld}$$

f_{cd}—deflection (initial) from a short-term load and is determined at $k_0 = 1$ and $\gamma_{bl} = 1$, i.e.

$$D_0 = E_b J_{x,red}$$

f_{ld} —deflection (retarded) from long-acting loads, $k_0 = 0,2$ and $\gamma_{bl} = 0,85$, that is,

$$D = K_b E_b J_{x,b} + k_s E_s J_{x,s}$$

The coefficients K_b and K_s are determined by formula (11).

Rebar (reinforcement)	**Classes**	**Ultimate stresses $\overline{\sigma_s}$, MPa**		
		Cracks are not harmful, R_s	**Cracks are dangerous**	**Cracks are very dangerous**
Round with a smooth surface η=1	A 240 A 300	210 260	140 150	105 110
Ribbed with a periodic profile η = 1,6	A 400 A 500	350 435	240 240	175 175
Welded nets: smooth wire surface η = 1	B 500	435	150	110
Ribbed with a periodic surface of the wire η = 1,6	B_p 500	435	240	180

Table 1.
The values of ultimate stresses in tensed reinforcement depending on the categories of crack resistance and the type of reinforcement surface η.

A detailed presentation of the design according to the second group of limit states is given in Ref. [8].

As a result of the development of a new method for designing reinforced concrete elements, the following conclusions can be drawn:

6. Conclusions

1. It has been established that the developed method for designing reinforced concrete elements, which is based on an accurate and rigorous application of the nonlinear deformation model of the mechanics of deformed solids to the problems of elastoplastic bending of the theory of reinforced concrete, on the basis of which the design of reinforced concrete structures will be carried out, fully meets the rules, principles, and requirements of the Eurocode.

   The essence of this technique lies in the fact that, under the strength conditions of reinforced concrete elements, expressed in terms of ultimate deformations of concrete of the compressed zone $\varepsilon_{b,ult}$ and tensed reinforcement $\varepsilon_{s,ult}$, and based on new concepts of limit states and the hypothesis of flat sections, they were reduced to compliance with new conditions written relative to the height of the compressed zone of concrete of bending elements, for example, at eccentric tension to compliance with the condition $0 \le \xi \le \xi_{OR}$, at simple bending to $\xi_{OR} \le \xi \le \xi_R$, and at eccentric compression to the following three conditions:

   $\xi_{OR} \le \xi \le \xi_R$, $\xi_R \le \xi \le 1$, and $1 \le \xi \le \infty$, where the height of the compressed zone ξ is determined from the general condition $M_x \le M_{x,ult}$. Ultimately, the calculation (design) method has a practical and accessible form.

2. The developed method for designing reinforced concrete elements makes it possible to take into account the main characteristics of materials: the long-term strength of concrete and the increase in the strength of reinforcement (rebar) after the yield point in accordance with the state diagrams of concrete and reinforcement.

3. Comparison of the limit states for the strength of reinforced concrete elements based on a nonlinear deformation model and a plastic hinge model shows that when choosing the limit states, it is necessary to proceed from the deformation diagram obtained on the basis of the hypothesis of flat sections and state diagrams of concrete and reinforcement, which leads to an expansion of the range of tasks that can be solved established by Eurocodes.

4. Given the above, it is proposed to bring the current national standards SNiP 52.01-2003 and AzDTN 2.16-1 for reinforced concrete in accordance with the Eurocodes, by reworking them on the basis of the developed methodology for calculating (designing) reinforced concrete structures.

Author details

Khanlar Seyfullaev
Azerbaijan Scientific-Research Institute of Construction and Architecture, Azerbaijan, Baku

*Address all correspondence to: xanlar.seyfullayev@mail.ru

References

[1] Construction Newspaper. Discrepancy between the Updated Standard for Reinforced Concrete and the Eurocode—An Obstacle in Construction. 19; 2014

[2] Bondarenko VM, Suvorkin DG. Reinforced Concrete and Stone Structures. Moscow: Textbook for Universities; 1987. p. 384

[3] Seyfullayev KK. Improvement of applications non-linear deformation model in calculation of bending reinforced concrete elements by new concept of limit states. Structural Mechanics and Design of Structures. 2020;**1**:52-61

[4] Seyfullayev KK, Garayev AN. About discrepancy between national standards for reinforced concrete and Eurocode. BST. 2017;**9**:40-45

[5] Seyfullayev KK, Garayev AN. Applications non-linear deformation model in calculation of reinforced concrete elements by new concept of limit states. Polish Journal of Science. 2018;**1**(1):34-47

[6] Seyfullayev KK, Garayev AN. Applications non-linear deformation model in calculation of bending reinforced concrete elements. Sciences of Europe. 2018 (Praha);**1**(33):51-60

[7] Seyfullaev KK. The method for calculating reinforced concrete elements based on a nonlinear deformation model of mechanics of deformable solids. Structural Mechanics and Design of Structures. 2021;**5**:64-73

[8] Pierre C. Calcul des ouvrages en beton armé suivat les régles BAEL—83. In: Théorie et aplication. Paris: Eyrolles; 1986. p. 460

[9] Beeby AW, Narayanan RS. Designer's Guide to Eurocode 2. Moscow: MGSU; 2013. p. 292

[10] Baykov VN, Sigalov E. Reinforced Concrete Structures Moscow. Stroyizdat; 1991. p. 767

[11] SniP 52-01-2003. Concrete and Reinforced Concrete Structures. Main Provisions, M, 2012

[12] AzDTN 2.16-1. Concrete and Reinforced Concrete Structures. Designing Standards, Baku 2015, p. 131

[13] Sakhnovskiy KV. Reinforced Concrete Structures. Moscow: Стройиидат 1951. 679 p

Chapter 11

Reinforced Concrete Design with Stainless Steel

Yakubu Mustapha Karkarna, Ali Bahadori-Jahromi, Hamid Zolghadr Jahromi, Emily Halliwell and Musab Mohammad Rabi

Abstract

In the design of reinforced concrete structures, the bond property is crucial. This is important for achieving the composite action between the two materials constituents, allowing loads to be efficiently transmitted. The higher strain hardening and ductility capacity of stainless steel over mild steel are one of its major benefits. International design codes, such as Eurocode 2, do not provide a separate design model for concrete structures with stainless reinforcing bars. The background paper to Eurocode 2 highlighted that there is no technical reason of why the Eurocode 2 design model cannot be used in conjunction with other types of reinforcement, provided allowance is made for their properties and behaviour. While this notion is valid when using a mild steel reinforcing bar, it produces erroneous results when a stainless reinforcing bar with a lap splice is used in a reinforced concrete section. Even though there has been a large number of studies on the behaviour of structure with stainless steel in recent years, most of it has been on plain stainless-steel members rather than reinforced concrete or stainless-steel reinforced concrete with lap splice. As a result, the purpose of this chapter is to evaluate and compare the behaviour of stainless and mild steel reinforced concrete with and without lap splices.

Keywords: mild steel, stainless steel, lap, splice length, reinforced concrete, cement-based materials

1. Introduction

The foundations of human civilisations have long been made of cement-based materials. These materials were changed in order to maintain their function in our lives as human activities advanced. Cement's main function is to serve as a hydraulic binder, strengthening the bond between fragmented particles so that they may be used in a variety of applications. The resulted material will differ from the initial materials in terms of its mechanical and physical properties. The exothermic hydration reactions that begin when the water and the binder are mixed are responsible for these changed properties.

In recent years, there has been an increased interest in and use of stainless-steel reinforcing bars in concrete buildings, due to its distinctive properties such as high ductility, long life cycle, excellent corrosion resistance and significant development of strain hardening. This chapter provides a comprehensive overview of prior studies on the design and behaviour of stainless steel as a structural material. The first section provides an overview of stainless steel in general, covering material properties, classifications, chemical composition, and grade categories. The second section delves into the details of using stainless steel as a reinforcing bar in concrete structures, including the obstacles and requirements. It also covers the bond performance of stainless steel in reinforced concrete. In addition, the current applications for stainless steel reinforcement are discussed in this section. The third section focuses on the design requirements of stainless-steel reinforcement outlining the distinctions between mild steel and stainless steel, particularly in terms of constitutive relationship, and discussing current codes of practice for concrete members reinforced with stainless steel.

1.1 Structural applications of stainless steel

In structural engineering, stainless steel is commonly employed for load-bearing applications mainly because of its superior corrosion resistance. It has good formability and recyclability, excellent mechanical characteristics, a long-life cycle, and requires very little maintenance [1]. When compared to mild steel, stainless steel has superior strain hardening capacity and ductility, making it ideal for use as a ductile section that warns of impending collapse. Stainless steel was first used in building in the 1920s for façade and roofing purposes [2, 3]. Stainless steels have recently gained popularity in load-bearing applications that require strength, ductility, durability and stiffness, as well as high resistance.

Stainless steels are manufactured in various forms including tube, plate, sheet, bar, fasteners and fixings, rolled and cold-formed structural sections. Because they are the most readily available and relatively simple to make, cold-formed sections manufactured from steel plates are the most widely utilised materials for structural components [4, 5].

1.1.1 Composition

Stainless steels are a class of corrosion-resistant alloying steels with a maximum carbon content of 1.2% and a minimum chromium concentration of 10.5% [6, 7]. Stainless steel's distinctive properties are determined by the constituent elements of stainless-steel alloy, therefore selecting the right grade for each purpose is essential. In all stainless-steel alloys, chromium is one of the most essential elements, because it offers corrosion resistance by forming a thin chromium oxide film on the material's surface in the presence of oxygen, leading to a passive protective layer [8, 9]. Other alloying elements that have a role in determining the characteristics of stainless steel are also essential. For instance, nitrogen significantly enhances the mechanical properties of the material, molybdenum improves the resistance against uniform and localised corrosion, and nickel improves the formability and ductility of the material [10]. Among the other alloying elements that are commonly present are: Sulphur, carbon, phosphorus, copper and silicon. The European Standard [6] provides comprehensive information on the chemical composition of various stainless-steel grades. The chemical composition of some commonly used stainless steel reinforcement grades is shown in **Table 1**.

1.1.2 Classification

Stainless steels are classified using a number of international categorisation systems. The American Iron and Steel Institute (AISI) specification and the European standards are the most extensively used. More details on these classification systems may be found in the sub-sections below.

1.1.2.1 The American iron and steel institute system

Stainless steels are classified by the AISI into different categories. Ferritic and austenitic stainless steels, for instance, are classed as 400 series alloys (e.g., 403, 409) and 300 series alloys (e.g., 316, 304). The fundamental flaw of this system is that it does not provide specifics on the chemical composition of each grade. **Table 1** lists some of the stainless-steel reinforcement grades available, along with their equivalent

1.1.2.2 European standard

The chemical composition of stainless steel is classified by the European standard [11, 12]. An individual number is assigned to correspond to the nominal alloy composition and then a generic number is assigned to each grade to identify it as part of a group. The numeral in grade 1.4436, for instance, represents:

- 1 represents the steel
- 44 represents the stainless-steel group
- 36 represents the individual material ID

American (AISI)	European (EN 10088-1)		Chemical composition (%)								
Grade	Name	Grade	C Max	Si Max	Mn Max	P Max	S Max	Cr Max	Ni Max	Mo Max	N Max
2205	X5CrNi 18-10	1.4301	0.3	1.0	2.0	0.35	0.015	21.0/ 23.0	4.5/ 6.5	2.5/ 3.5	0.10/ 0.22
2304	X5CrNiMo 17-12-2	1.4401	0.3	1.0	2.0	0.35	0.015	22.0/ 24.0	3.5/5.5	0.1/ 0.6	0.05/ 0.20
LDX 2191	X2CrNiMoN 17-13-3	1.4429	0.3	0.4	5.0	—	—	21.5	1.5	0.3	max 0.22
316LN	X3CrNiMoN 22-2-0	1.4162	0.3	1.0	2.0	0.045	0.015	16.5/ 18.5	11.0/ 14.0	2.5/ 3.0	0.12/ 0.22
316	X3CrNiMoN 23-4	1.4362	0.7	1.0	2.0	0.045	0.3	16.5/ 18.5	10.0/ 13.0	2.0/ 2.5	Max 0.11
304	X2CrNiMoN 22-5-3	1.4462	0.7	1.0	2.0	0.045	0.3	17.5/ 19.5	8.0/ 10.5	—	Max 0.11

Table 1.
The chemical composition of various stainless-steel grades [10].

To provide more information about a grade's chemical composition, the grade number is also given the corresponding grade name. For instance, grade 1.4436 is designated as X3CrNiMo 17-13-3, which means:

- X represents a high alloy steel
- 3 represents percentage of carbon content
- CrNiMo is the chemical symbol of the main alloying elements.
- 17-13-3 refers to the nominal percentage of the main alloying elements

1.1.3 Categories of stainless steel

There are five primary types of stainless steel that are categorised based on metallurgical structure, namely, precipitation hardened, duplex, austenitic, martensitic and ferritic grade. In structural applications, such as stainless-steel reinforcement, duplex and austenitic grades are most common. An overview of each category's primary benefits and drawbacks are as follows:

1.1.3.1 Precipitation hardened stainless steel

Precipitation hardened stainless steel offers superior corrosion resistance over martensitic or ferritic stainless steel and is identical to austenitic grades, which include 8% nickel and 18% chromium [13, 14]. Depending on the heat treatment conditions, they have good ductility, toughness and strength. These grades are primarily utilised in aerospace and oil and gas industries, and construction industries for applications like tie-bolts [15].

1.1.3.2 Duplex stainless steel

Due to the mixed microstructure of austenitic and ferritic stainless steels, duplex stainless steels are also referred to as austenitic-ferritic stainless steels. Duplex stainless steels typically include 4–5% nickel and 22–23% chromium [3]. These grades show great ductility and high strength properties as well as excellent corrosion resistance. Duplex stainless steel should be utilised where the materials are exposed to a contaminated or aggressive environment or high strength is required. As a result, they are extensively utilised as shafts, tension bars, valves and pin connections in offshore structures and chemical industries [13, 16].

1.1.3.3 Austenitic stainless steel

Due to their high corrosion resistance and excellent to mechanical properties, austenitic stainless steels are the most frequently utilised grades in structural applications. They generally include at least 8–11% nickel and 17–18% chromium [4, 17]. The austenitic grades provide excellent formability and weldability as well as a wide range of service temperatures [17, 18]. Austenitic stainless steels have been employed in industrial piping, housewares, architectural facades, containers and load-bearing structural members.

1.1.3.4 Martensitic stainless steel

The carbon content of martensitic stainless steel is greater than that of other grades. Martensitic stainless steels have a microstructure identical to carbon steels and ferritic stainless steels. These grades have corrosion resistance that is similar to ferritic grades. In comparison with ferritic, austenitic and duplex grades, they have a lower ductility [18, 19]. Moreover, heat treatment is also required for martensitic stainless steels before and after welding. Despite their lower cost in comparison to other stainless-steel grades, their weldability requirements and poor corrosion resistance limit their applicability to knife blades and valves. They are not utilised in load-bearing applications.

1.1.3.5 Ferritic stainless steel

The chromium concentration in ferritic stainless steel is normally between 11% and 17% [20]. They have an atomic structure that is similar to carbon steel and contain less nickel than austenitic stainless steel. As a result, they have usually limited toughness and ductility and poorer weldability, formability and corrosion resistance than austenitic stainless steels. Because of the limited nickel content, ferritic stainless steels are less costly and have less price volatility. Ferritic stainless steels are better suited to interior applications like handrails and shop fittings, as well as other domestic items like boilers and washing machine parts because of their less corrosion resistance compared to other grades [15].

1.1.4 Material properties

Stainless steels provide excellent corrosion resistance as well as weldability, toughness and strength. Stainless steel material properties differ based on a number of factors, including the direction of rolling and level of cold-working, material thickness and chemical composition. When compared to carbon steels, duplex and austenitic stainless steels offer significant strain hardening properties and higher strength. These grades are known for their high ductility, which may exceed 40%. Stainless steels that are martensitic or ferritic have lower strain hardening and strength. Precipitation-hardened stainless steels, on the other hand, have exceptionally high strength, often exceeding 1500 N/mm^2, although they have limited ductility, depending on the heat treatment condition [20]. The elasticity modulus of various stainless-steel categories is identical to that of carbon steel in general. A value of 200,000 N/mm^2 may be used to define the elasticity modulus for all stainless-steel grades, based on the European standard [6]. **Table 2** presents information on the mechanical properties of some typical grades of stainless steel.

1.1.5 Recycle

A large amount of waste material is generated by the construction industry. The use of more environmentally friendly materials is required in the construction industry to minimise waste. Stainless steels, in this regard, are long-lasting materials with a high residual value of fundamental elements such as molybdenum, chromium and nickel [20]. Around 80% of new stainless steel manufactured in Europe is created from recycled waste stainless steel, according to research [21]. This gives stainless steel more environmental and economic benefits.

Stainless steel type	Grade	Minimum 0.2% proof strength (N/mm^2)	Ultimate tensile strength (N/mm^2)	Modulus of elasticity, E (kN/mm^2)	Minimum elongation after fracture (%)
Ferritic	1.400 (410S)	220	400–600	200	19
	1.4512 (409)	210	380–560	200	21
Duplex	1.4362 (SAF2304)	400	600–850	200	20
	1.4462 (2205)	400	640–840	200	20
Austenitic	1.4301 (304)	210	520–720	200	45
	1.4307 (302L)	200	500–650	200	45
	1.4401 (316)	220	520–670	200	40
	1.4404 (316L)	220	520–670	200	40

Table 2.
Mechanical properties of stainless steels grades [6].

1.1.6 Cost

Stainless steels are inevitably more costly than carbon steels in structural applications [4, 11]. This prevents stainless steel from becoming more widely used. Nevertheless, the initial material cost, on the other hand, does not reflect the overall cost of the construction during its lifetime. Other aspects such as inspection and maintenance costs as well as the immediate cost associated with fire and corrosion protection must be taken into account in order to make an informed decision. When all of these aspects are considered simultaneously, stainless steel is a superior option to carbon steel, particularly for buildings that are subjected to extreme environments.

1.2 Stainless steel reinforcing bar in concrete structures

One of the most commonly used structural solutions in building construction is reinforced concrete. It is popular because it is an efficient, cost-effective and versatile solution with plenty of performance criteria and design guidelines. Owing to their great ductility, significant strain hardening, excellent durability, exceptional corrosion resistance and long-life cycle, stainless steel has recently been used in reinforced concrete structures. Because of the effective usage of readily available constituent materials, reinforced concrete constructions are extensively employed for a variety of applications such as bridges, multi-storey buildings and tunnels.

The stainless steel's constitutive behaviour differs significantly from that of carbon steel as it shows a rounded behaviour from the start, with high ductility and significant stain hardening and without a clearly defined yield point. As shown in **Figure 1**, carbon steel has a more linear relationship in the elastic stage with a moderate degree

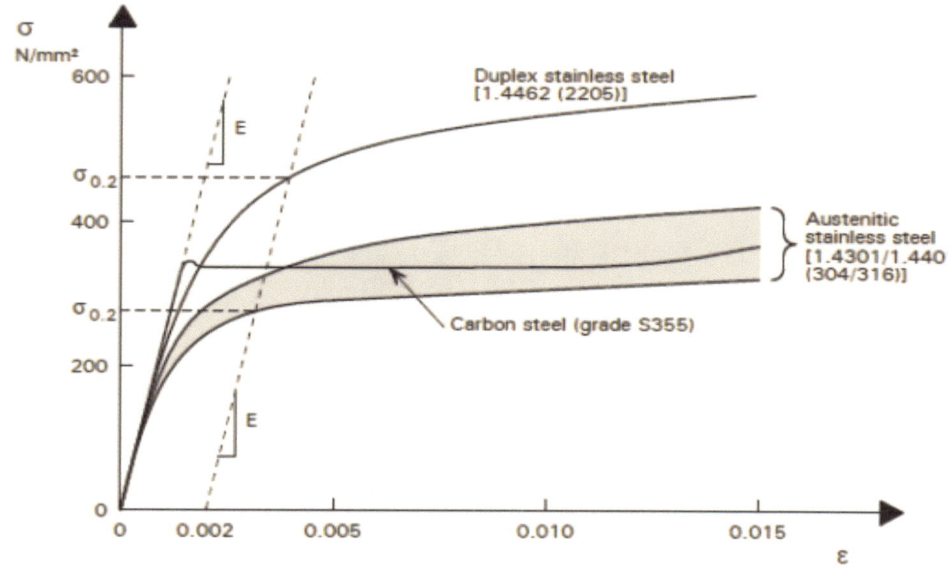

Figure 1.
Stress–strain curves for stainless steel and carbon steel [15].

of strain hardening and a well-defined yield point. When there is no observable yield point, 0.2% proof stress is typically used in the design.

Stainless steels are often represented using the modified-Osgood stainless steel material model which is an improvement of the original version presented in [8].

1.2.1 Life cycle cost

Stainless steel reinforcing bars have a higher initial cost than ordinary carbon steel reinforcing bars, ranging from 3 to 8 times more depending on the grade [13, 22]. Due to the high initial cost, stainless steel reinforcement is sometimes restricted to the outer layer, which is more vulnerable to chloride-ingress. Despite this, stainless steel reinforcement has been proven to save total maintenance costs by up 50% throughout the life of a structure, particularly marine and bridge structures [5, 23]. According to a study conducted by the Arup Research and Development team and oversaw by the UK Highways Agency, stainless steel reinforcement may drastically enhance the lifetime of buildings while simultaneously lowering maintenance costs [24]. Incorporating stainless steel in concrete structures reduces the amount of rehabilitation and maintenance work required during their lifetime. These qualities are critical for infrastructure and highways to minimise rerouting and road closures as well as carbon emissions and delays that come with them. Furthermore, employing corrosion-resistant reinforcing bars like stainless steel saves a lot of money by allowing some durability criteria such as the need for reinforcement coating, design crack width and depth of concrete cover to be relaxed. Incorporating these adjustments into the design of reinforced concrete buildings might save a lot of money, especially on big projects. The Oland Bridge in Sweden, which employed both carbon and stainless-steel reinforcing bar is presented in **Figure 2** as an instance of real-life cycle costs. The data in the figure demonstrated that the cost of a bridge with stainless steel stays unchanged during its lifetime, suggesting no extra expenses, but the cost of a carbon

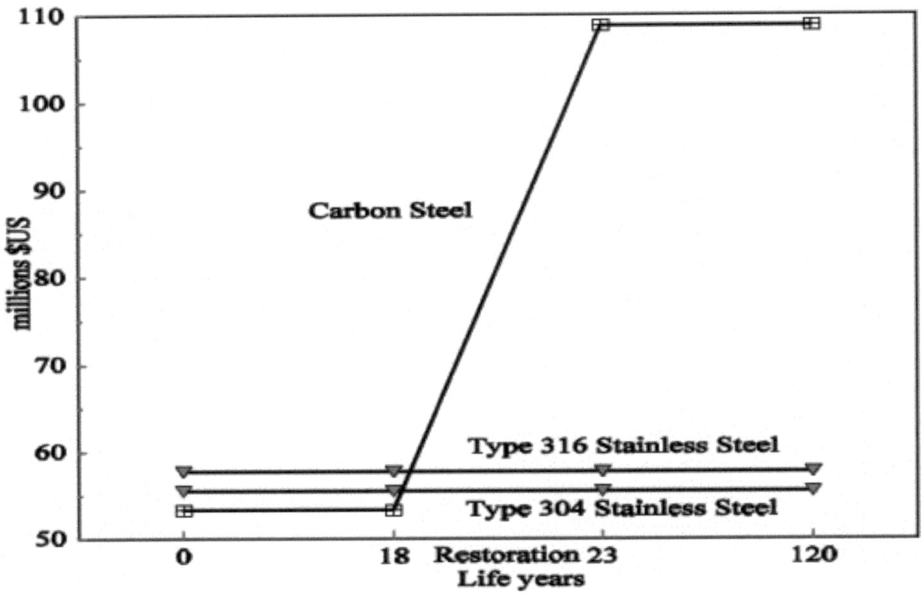

Figure 2.
Life cycle cost analysis for Sweden's Oland bridge [25].

steel-reinforced concrete solution increases drastically after around 20 years. Another research on the Schaffhausen bridge in Switzerland found that stainless steel grade has a 14% lower life cycle cost than carbon steel rebar [25]. This is compelling proof of stainless-steel reinforcement's long-term cost-effectiveness in infrastructure projects.

1.2.2 Durability

Due to the high maintenance costs related to carbonation and steel reinforcement corrosion and of the concrete, there is a growing desire to increase the life cycle cost and durability of reinforced concrete buildings. This is especially true for buildings in harsh environments like those found in industrial or coastal and marine settings. Corrosion is hard to avoid in buildings with carbon steel and exposed to a harsh environment. Changing some design parameters, such as controlling the alkalinity of the concrete mix or the thickness of the concrete cover, is a common approach to increasing the durability of reinforced concrete structures [26]. However, in harsh environments, these precautions may not be sufficient to stop the intolerable level of corrosion from forming. In this respect, the utilisation of stainless-steel reinforcement in exposed structures such as tunnels, bridges and retaining walls can be an effective way to combat corrosion and deterioration. This may even mean that the structure will not require rehabilitation works and expensive inspection in the future. Existing concrete structures can also be rehabilitated and restored using stainless steel reinforcement [16, 27].

1.2.3 Mechanical behaviour

When compared to traditional carbon steel, stainless steel reinforcement has a superior mechanical behaviour. In recent years, there has been a small number of experiments on the mechanical behaviour of stainless-steel reinforcement. When compared to carbon steel reinforcement, austenitic stainless steel reinforcement grades 1.4429 and 1.4311 give superior hardness and strength properties [28]. Reference [29] studied the mechanical and ductility properties of duplex and austenitic stainless steel reinforcement grades 1.4482, 1.4301 and 1.4362 with reference to carbon steel grade B500SD. It was discovered that stainless steels have three times the ductility of carbon steel. However, compared to carbon steels, these stainless steels had a 15% lower elasticity modulus. This is attributed to the fact that stainless steels show nonlinear behaviour from the beginning, making the modulus of elasticity difficult to measure. **Table 3** shows an overview of some of the mechanical parameters of several grades of stainless-steel reinforcement. The stainless steels have great ductility, significant strain hardening and excellent tensile strength. In order to minimise unexpected collapse, these characteristics are very essential in design.

1.2.4 Commonly used stainless steel reinforcement grade

In the open market, stainless steel reinforcement is available in a variety of grades, including duplex grades 1.4362, 1.4162 and 1.4462, as well as austenitic grades 1.4301, 1.4307 and 1.4311. Grade 1.4307 is a standard low-carbon austenitic stainless steel and the most commonly found grade used in construction, whereas grade 1.4311 is a low-carbon austenitic stainless with improved strength and low-temperature toughness due to its higher nitrogen and nickel content. Both of these grades are appropriate for structural applications that require minimal magnetic strength. Due to the relatively high nickel content compared to the austenitic grades, grade 1.4362 is duplex stainless steel that offers excellent corrosion resistance. The lean duplex grades are a new form of duplex stainless steel that has been produced in recent years and has a comparatively low nickel content. Due to the low nickel content, grade 1.4162 offers exceptional corrosion resistance while also having nearly twice the characteristic strength of austenitic steel for almost the same cost.

1.2.5 Selection and classification of stainless-steel reinforcement

The excellent corrosion resistance of stainless-steel reinforcement is undoubtedly one of the most important benefits. As result, categorising stainless steels based on

Product form	Grade	Bar diameter (mm)	Yield strength $\sigma_{0.2}$ (N/mm^2)	Tensile strength (N/mm^2)	Modulus E (kN/mm^2)	Elongation ε_u
Ribbed bars	1.4311	12	480	764	202.6	48.3
	1.4311	16	528	717	199.9	47.9
	1.4162	12	682	874	199.1	32.4
	1.4162	16	646	844	195.2	32.9
	1.4362	16	608	834	171.4	35.1
Plain round bars	1.4307	12	562	796	210.2	39.9
	1.4307	16	537	751	211.1	42.4
	1.4162	12	805	964	308.7	18.8
	1.4162	16	760	860	197.5	22.0

Table 3.
Mechanical properties of stainless-steel reinforcement [23].

their corrosion resistance may make choosing the appropriate grade easier. The Pitting Resistance Equivalent Number (PREN) is the most commonly used categorisation technique for measuring the relative corrosion resistance of a metal. The content of molybdenum, chromium and nickel in an alloy determines the corrosion resistance of the metal, as indicated in Eqs. (1) and (2) [10].

$$\text{PREN} = \%\text{Cr} + 3.3\ (\%\text{Mo}) + 30\ (\%\text{N}) \quad \text{for duplex stainless steels} \tag{1}$$

$$\text{PREN} = \%\text{Cr} + 3.3\ (\%\text{Mo}) + 16\ (\%\text{N}) \quad \text{for austenitic stainless steels} \tag{2}$$

When it comes to choosing the appropriate grade of stainless steel for a specific application, classifying stainless steel by its PREN number is useful. However, the PREN ignores the beneficial effects that come from the concrete cover and also does not take into account the chloride threshold of each grade on the passivity of stainless steel [10].

Table 4 shows a categorisation example for stainless steel reinforcement. In this example, the reinforcement is divided into four categories based on their PREN, the surrounding environment and the lifetime of the structure. Class 0 is proposed for structures with a design service life of 10–30 years that are subjected to relative humidity and moderate temperature and are located in the marine environment. For the same conditions, class 1 is recommended with a design service life of 50–100 years. Class 2 is appropriate for structures with a moderate design service life and high chloride penetration, as well as moderate to high relative humidity and temperature. Finally, class 3 is suggested for marine environment structures that required a long design service life in relative humidity and high temperature.

Table 5 gives suggestions for choosing the suitable grade of stainless-steel reinforcement according to the exposure conditions as recommended by the design manual for bridges and roads [31] for infrastructure and highways.

1.2.6 Use of stainless-steel reinforcement

It is widely acknowledged that carbon steel reinforcement in reinforced concrete structures may not be as durable as previously anticipated in all conditions [12, 18]. Corrosion of carbon steel reinforcement in harsh environments like coastal and marine regions can lead to inconvenient rehabilitation, challenging and very expensive work. Stainless steel reinforcement is an effective and long-lasting option in this situation. The Progresso Pier in Mexico depicted in **Figure 3**, was built in the 1940s employing grade 1.4301 austenitic stainless steel and is one of the earliest examples of the usage of

Corrosion resistance class	Steel type	Stainless steel grade	PREN
Class 0	Carbon steel	N/A	N/A
Class 1	Austenitic stainless steel (without molybdenum)	1.4542	17
		1.4301	19
Class 2	Austenitic stainless steel (with molybdenum)	1.4571	25
		1.4436	26
		1.4429	26
		1.4401	25
Class 3	Duplex	1.4462	36

Table 4.
Stainless steel reinforcement categorisation based on their corrosion resistance [10].

Exposures condition	Stainless steel grade
Specific structural requirements for the use of higher strength reinforcement and suitable for all exposures	1.4429 1.4462
Stainless steel reinforcement embedded in concrete with normal exposures to chlorides in soffits, diaphragm walls, edge beams, substructures and joints	1.4301
Direct exposures to chlorides and chloride-bearing waters for example dowel bars, holding down bolts and other components protruding from the concrete.	1.4429
As above but where additional relaxation of design for durability is required for specific reasons on a given structure or component that is where waterproofing integrity cannot be guaranteed over the whole life of the structures.	1.4436

Table 5.
Stainless steel grade selection [30].

stainless-steel reinforcement. The bridge has been in service for more than 70 years without requiring any significant maintenance or major repair work.

Other projects that have utilised stainless steel reinforcement include the Sheikh Zayed Bridge in Abu Dhabi and Stonecutters Bridge in Hong Kong, as depicted in **Figures 4** and **5**, respectively. These two bridges are made with duplex stainless-steel grade 1.4462. The stainless-steel reinforcing bars are well situated and only utilised for the reinforcement outer layer in both bridges in the supposed splash zone. The Broadmeadow Bridge in Ireland and the Highnam bridge expansion project in the UK both employed grade 1.4436 stainless steel. The Queensferry Crossing in Scotland, which opened in 2017, is one of the high-profile and recent applications of stainless-steel reinforcing bars. Stainless steel reinforcement has been employed for restoration and renovation as well as new construction. The pillars and stone arches of the Knucklas rail bridge, for instance, were rehabilitated using austenitic grade 1.4301 stainless steel reinforcement [25].

Figure 3.
The Progresso Pier in Mexico [32].

Figure 4.
Stonecutters bridge [13].

Figure 5.
Sheik Zayed bridge [33].

1.2.7 Fire behaviour

One of the most crucial attributes for creating fire-resistant buildings is the material's capacity to maintain strength and stiffness at high temperatures. Because of the chemical composition, stainless steel has exceptional strength and stiffness retention at extreme temperatures [2]. There have been a lot of studies into stainless steel fire performance [27, 34–36] but very little research studies into the stainless-steel performance at high temperatures [23].

Figures 6 and 7 show a comparison of stiffness and strength retention factors between carbon steel and stainless steel at 0.2% proof stress. In comparison to carbon steel, stainless steel has a distinct advantage in terms of stiffness and strength at high temperatures. In the event of a fire, these distinguishing characteristics are immensely useful and give the structure the resistance needed for a longer duration of time.

1.2.8 Corrosion behaviour

Corrosion of the reinforcement, particularly for members subjected to a harsh environment, is now recognised as one of the most significant issues faced by reinforced concrete structures [37]. Corrosion is a major issue that causes weakness in the bond strength between the surrounding concrete and the reinforcement, as well as the reduction in the nominal reinforcement area which affects the integrity and safety of concrete structures. Corrosion takes place due to carbonation and chloride penetration of concrete. While the former is caused by carbon dioxide in the surrounding air attacking the calcium in the concrete. While Ingress of chloride from the marine environment or de-icing salts in frosty weather causes the latter. The corrosion protection of the reinforcement in a typical reinforced concrete design is mostly dependent on the durability of the steel passivation layer and the concrete cover. This passivation layer on typical carbon steels can quickly break down, allowing corrosion to form, particularly in a hash or contaminated environment.

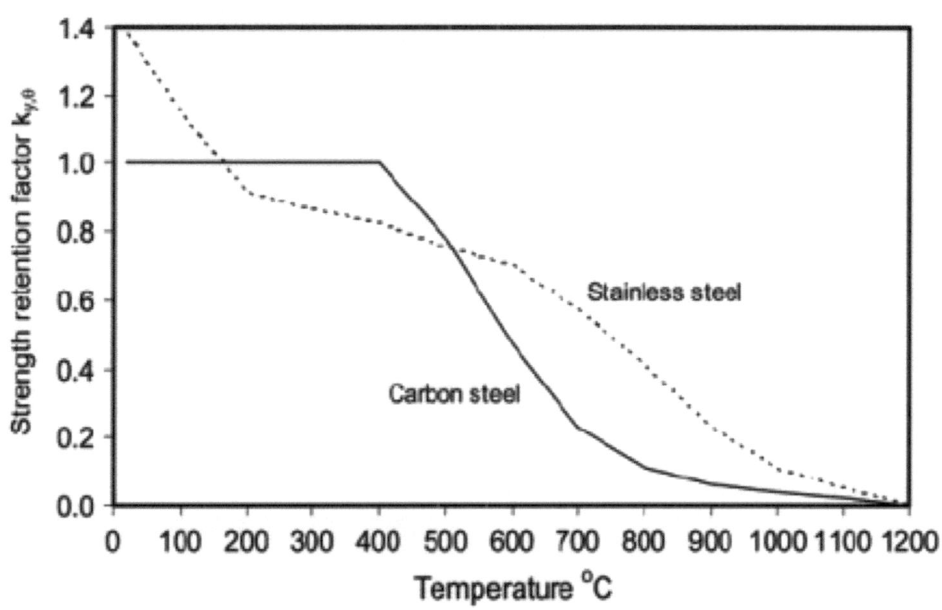

Figure 6.
Comparison of carbon steel and stainless-steel strength retention factor [2].

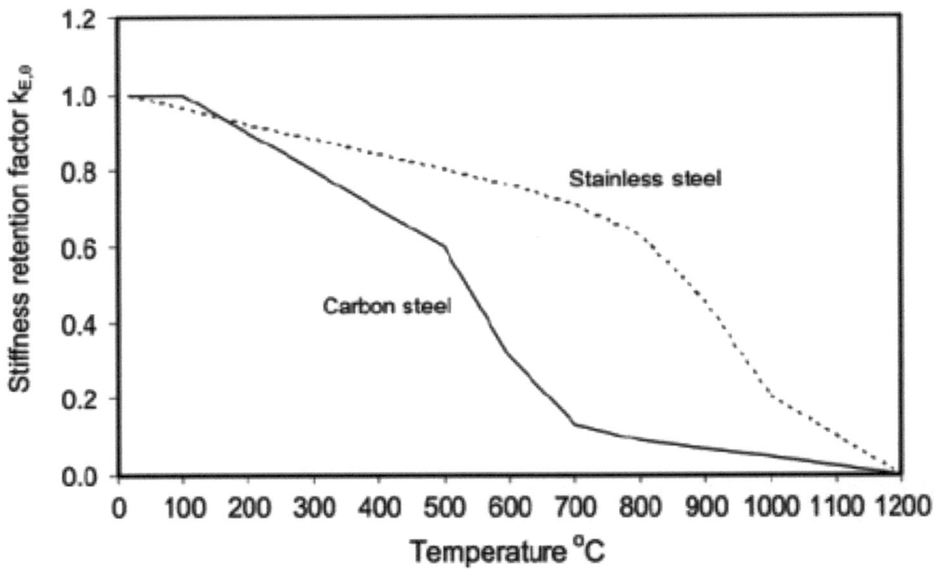

Figure 7.
Comparison of carbon steel and stainless-steel stiffness retention factor [2].

The usual methods of decreasing the potential corrosion risk are to control the concrete alkalinity, use reinforcement coating materials or cement inhibitors or increase the depth of concrete cover [20]. These precautions, however, may not be sufficient to avoid corrosion to undesirable levels. In this situation, stainless steel reinforcement is an excellent alternative for dealing with inherent corrosion issues. Due to its high chromium content (i.e., a minimum of 10.5%), stainless steel offers excellent corrosion resistance even in a harsh environment. In the presence of oxygen, chromium produces a thin self-regenerating chromium oxide coating on the material's surface forming a strong passive protective layer [8, 19, 35].

Austenitic stainless-steel reinforcement is 10 times more corrosion resistant than carbon steel reinforcement [38, 39]. When compared to austenitic reinforcement, duplex reinforcement has equivalent or even greater corrosion resistance [32, 39, 40]. The corrosion performance of several grades of stainless is compared to that of carbon steel in **Figure 8**. The x-axis depicts the influence of concrete's PH value, while the y-axis depicts the effect of chloride concentration. Even at very low chloride contents, it is clear that carbon steel has poor corrosion resistance. The PH value of carbon steel is also very sensitive as corrosion occurred. Stainless steel reinforcement, on the other hand, has great corrosion resistance even at low PH values and high chloride content.

1.2.9 Bond behaviour

In the design of reinforced concrete structures, bond is an essential property. It is necessary to ensure that the composite action between the two materials constituent is attained, allowing loads to be transferred efficiently. Insufficient concrete-steel bond can cause excessive rotation or deflection, ineffective anchorage of the reinforcing bar, as well as excessive slippage of the reinforcement leading to serious cracking of the concrete. The many interconnected parameters that determine the development of a bond make it a complicated phenomenon. The surface geometry of the reinforcing bar and the quality of

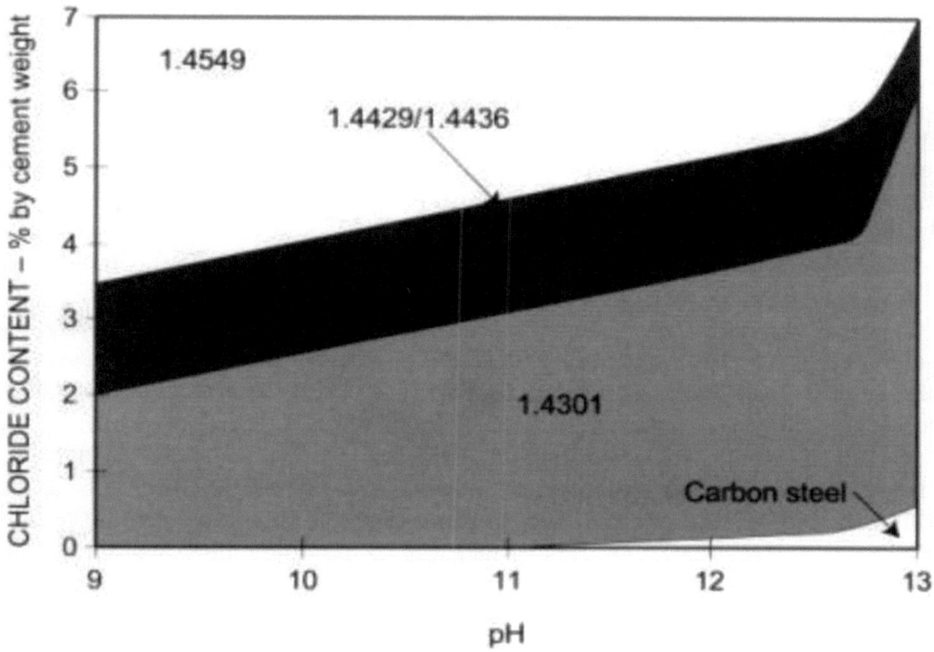

Figure 8.
Corrosion performance of various stainless steel compared to carbon steel [41].

the concrete are the most important parameters. Other key parameters are bar size, the cover distance, the direction of casting with respect to the orientation of the bars, clear space between adjacent bars and the number of reinforcement layers.

There has been little investigation into the bond performance of stainless-steel reinforcement in corrosive environments [20, 42]. And even a few research on the bond behaviour of stainless-steel reinforcement in normal conditions, with some research indicating that the bond developed by some duplex and austenitic stainless-steel bars is relatively low in comparison to similar carbon steel reinforcing bars [34, 43]. As a result, more study into the bonding properties of stainless-steel reinforcing bars in concrete is necessary.

Due to many international design guidelines, such as [44, 45] do not have specific bond design rules for stainless steel-reinforced concrete structures, designers generally use the same design guidelines established for conventional carbon steel reinforcing bars when designing reinforced concrete structures with stainless steel reinforcement. Because stainless steel has been reported to have a reduced bond strength than carbon steel, this is not always a safe approach unless particular test data is presented.

1.3 Current design models for laps

Design models for required lap length are used in design codes for structural members to account for stress formed in bond regions. With each new code reissue, the design models are updated regularly [45], which has been in use since 2004, and includes a lap design model that was originally published in the Ref. [44]. For the next generation of Eurocode 2, the project team for Eurocode 2 has developed a new proposal [46]. The lap design model was derived from [47], which is the background document for ref. [44]. The Eurocode 2 project team provides preliminary calibration factors for change from average values provided in [47] to design values that are still to be verified [46] for lap design models.

Models with and without bond strength definitions are differentiated in code provisions for laps. Earlier design codes specified lap lengths for different concrete classes based on bond strength (for example, [32, 44]). The ACI and Fib Bulletin models, on the other hand, were developed from statistical analysis of experimental data that took into consideration the maximum bar strength in laps without determining bond strength. The bond strength for Model Code 2010 was determined from the [47] design model. Bond strength is an optional parameter that simplifies the design, but it is not required for lap design.

Experimental and partially computational studies with finite element analysis are used to develop models for estimating developable stresses in laps. The processes for finding calibration factors, the influencing parameters, and the composition of equations differ. The major impacting parameters for all the models are compressive strength and lap length, however, their components differ. Some models introduce a summand for shear link contribution; while others consider the shear link contribution by a coefficient. A cover-to-bar diameter ratio is used in all models to account for the influence of concrete cover. The models take into account the influence of shear links differently. Some models comprise the transverse bar spacings_{st} , while others consider the number of shear link bars n_{st}. The correlation is provided by expression (3).

$$\sum A_{st} = n_{st} A_{st} = \left(\frac{l_0}{s_{st}} + 1\right) A_{st} \approx \frac{A_{st} l_0}{s_{st}} \tag{3}$$

The horizontal cracking plane is crossed by the number of shear link leg in side splitting. As a result, if side splitting is assumed, the shear link leg number n_l is included in the lap design models. The shear link cross-section is only evaluated once in face splitting design models because it only provides tensile resistance in one face crack. The number of shear link leg is not considered in this case (see **Figure 9**).

The design expression for laps based on Eurocode, National Annex, and ACI are provided in this section. The origin of the design expressions in the proposal for the next generation of Eurocode 2 from ref. [47] and in ref. [44] are discussed.

1.3.1 Fib bulletin 72

Reference [47] describes the background of bond strength established in ref. [44]. The semi-empirical expression for estimating the average bar stress in tension lap joints was obtained from 800 tests performed in Asia, Europe and the United States.

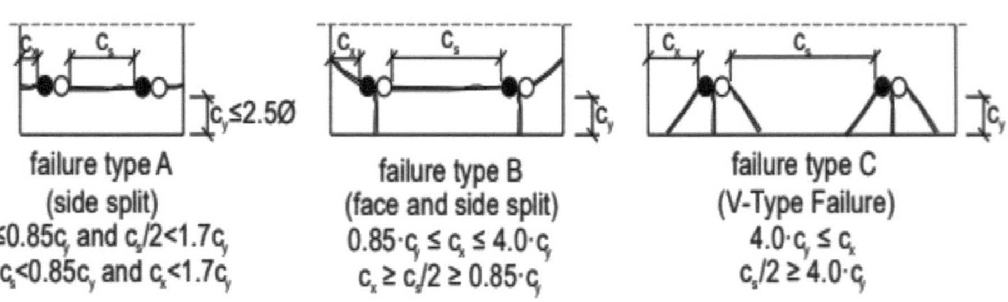

Figure 9.
Splitting failure modes. (Reproduced from [48] based on [49]).

The expression for average lap stress representing the authenticated affecting factors is provided as follows:

$$f_{stm} = 54(f_{cm}/25)^{0.25}.(25/\varnothing)^{0.2}.(l_b/\varnothing)^{0.55}.\left[(c_{min}/\varnothing)^{0.25}.(c_{max}/c_{min})^{0.1}.\ k_m k_{tr}\right] \leq f_y \tag{4}$$

Where;
f_{cm} is the measured concrete cylinder compressive strength
c_{min} is the minimum cover concrete
k_{m} is the coefficient of efficiency of shear link
c_{max} is the maximum cover concrete
l_{b} is the lap or anchorage
k_{tr} is the density of shear link

$$k_{\text{tr}} = n_l n_{st} A_{st}/(n_b \varnothing l_b) \leq 0.05$$

n_{b} is the number of lapped bars at a section
n_{st} is the number of stirrups in the lap length
f_{y} is the yield stress
n_{l} is the number of stirrups legs that crosses the potential splitting failure plane

The bond strength does not increase with a shear link ratio k_{tr} above 0.05. The parameter k_{m} compensates for the efficacy of the shear link depending on possible failure planes and their position. The shear link is particularly effective when the lap or an anchored bar has a lesser spacing to the next shear link leg across a splitting crack. When the horizontal spacing between bars is greater than 5Ø or 125 mm, the efficiency is decreased by 50%. There is zero effect on bond strength if the shear link does not cross the splitting carack.

Since test results outside of these boundaries hardly exist, Eq. (4) is restricted to the following boundary conditions.

- Good bond condition

$$\frac{l_0}{\varnothing} \geq 10$$

$$\frac{15}{\varnothing} \leq 5$$

$$\frac{25}{\varnothing} \leq 2$$

$$\frac{c_{max}}{c_{min}} \leq 5$$

$$0.5 \leq c_{min}/\varnothing \leq 3.5$$

The stress developed by bond increases when the transverse pressure is present to

$$f_{stm,tr} = f_{stm} + 6(l_b/\varnothing)p_{tr} < 1.75 f_{st,0} + 0.8(l_b/\varnothing)p_{tr} < 8.0(l_b/\varnothing) f_{cm}^{0.5} \tag{5}$$

Where:

$f_{st,0}$ average stress formed by bond for the base conditions of confinement with

$$f_{st,0} = 54(f_{cm}/25)^{0.25}(l_b/\varnothing)^{0.55}(25/\varnothing)^{0.2} \tag{6}$$

p_{tr} average compression stress perpendicular to the potential splitting failure surface
f_{stm} is the mean estimated stress developed in the bar

1.3.2 Model code 2010

The International Federation for Structural Concrete (known as the Federation Internationale du Beton or Fib) offers advice for the design of prestressed and reinforced concrete in the ref. [44]. In order to establish the required lap length, Model Code 2010 like Eurocode 2, necessitates the calculation of the design bond strength (f_{bd}).

The design bond strength f_{bd} provided in Model Code 2010 was determined by rewriting ACI express for transverse reinforcement index Eq. (34) with a lead coefficient of 41 to enable the formation of the reinforcement design strength $f_{yd} = 435$ MPa. The basic bond strength $f_{bk,0}$ was determined by rearrangement of Eq. (4) with a coefficient of 41. The shear link stress f_{stk} was set to $\frac{500}{1.5} = 435$ MPa and

$$l_{b,0}/\varnothing = \left(f_{yk}/\gamma_c.41\right)^{1.82}.(f_{cm}/25)^{-0.45}.(25/\varnothing)^{-0.36} = 73.5.(f_{cm}/25)^{-0.45}.(25/\varnothing)^{-0.36} \tag{7}$$

$$f_{bk,0} = f_{yd}.\varnothing/4.l_{b,0} = 1.5.(f_{cm}/25)^{0.45}.(25/\varnothing)^{0.36} \tag{8}$$

The coefficient, as well as the indices, were approximated to more practical values, and the confining reinforcement and cover values equivalent to the least detailing requirements were established. The coefficient 1.5 was modified to $\eta_1 = 1.6$ for the calculation of the basic bond strength without stirrups. The coefficient η_1 was modified to 1.75 and the values of basic bond strength were increased by 10% to account for the increase in bond strength if minimal confining reinforcement is provided [44]. The design bond strength is calculated using Eq. (9).

$$f_{bd} = (\alpha_2 + \alpha_3)f_{bd,0} - 2P_{tr}/\gamma_{cb} < 2f_{bd,0} - 0.4P_{tr}/\gamma_{cb} < 1.5\sqrt{f_{ck}}/\gamma_{cb} \tag{9}$$

Where:

$f_{bd,0}$ is the basic bond strength, which is derived using Eq. (10) and is a function of the characteristic compressive strength (f_c)

$$f_{bd,0} = \eta_1\eta_2\eta_3\eta_4\left(\frac{f_c}{25}\right)^{0.5} \tag{10}$$

η_1 is a coefficient taken 1.75 for ribbed bars (including stainless and galvanised reinforcement)

η_2 represents the casting position of the bar during concreting: $\eta_2 = 1.0$ for good bond condition.

γ_{cb} is the partial safety factor for bond $\gamma_{cb} = 1.5$

The influence of passive confinement from transverse reinforcement and concrete cover is represented as α_2. and α_3. Where; $\alpha_2 = \left(\frac{C_{min}}{\varnothing}\right)^{0.5} \cdot \left(\frac{C_{max}}{C_{min}}\right)^{0.15}$ and $\alpha_3 = k_d.(k_{tr} - \alpha_t/50) \geq 0.0, k_{tr} \leq 0.05$

Where:

C_{min} is the minimum cover concrete: $c_{min} = \min\left\{c_x; c_y; \frac{c_s}{2}\right\}$

C_{max} is the maximum cover concrete: $c_{max} = \min\left\{c_x; c_y; \frac{c_s}{2}\right\}$

α_t is the coefficient for the bar diameter;

$\alpha_t = 1.0$ for $\varnothing = 25$ mm

$\alpha_t = 0.5$ for $\varnothing = 50$ mm

$k_{tr} = \mathfrak{n}_t.\frac{A_{st}}{(\mathfrak{n}_b \varnothing s_t)}$ is the density of transverse reinforcement, relative to the lapped bars;

$\mathfrak{n}_l$ is the number of legs of confining reinforcement crossing a potential splitting failure surface at a section;

A_{st} is the cross-sectional area of one leg of a confining bar (mm^2);

s_t is the longitudinal spacing of confining reinforcement (mm);

$\varnothing$ is the bar diameter;

$\mathfrak{n}_b$ is the number of pairs of lapped bars in the potential splitting failure section;

$\mathfrak{n}_3$ represent the bar diameter: $\mathfrak{n}_3 = 1.0$ mm for $\varnothing \leq 25$

$\mathfrak{n}_4$ represents the characteristics strength of steel reinforcement that is being lapped (see **Table 6**).

k_d is an effective factor dependent on the reinforcement details and accounts for the stress developed and the nonlinear behaviour between the lap length in the bar (see **Figure 10**).

In case the concrete class is grade C60 or below and the anchored bar's diameter is less than 20 mm, the stirrup added for other reasons can be deemed to be adequate to meet the least criteria for confined reinforcement without additional explanation [44]. Therefore, the minimum stirrup has to be located with:

$\mathfrak{n}_4$	Characteristic strength of steel reinforcement f_{yk} (MPa)
1.2	400
1.0	500
0.85	600
0.75	700
0.68	800

Table 6.
Coefficient $\mathfrak{n}_4$ [50].

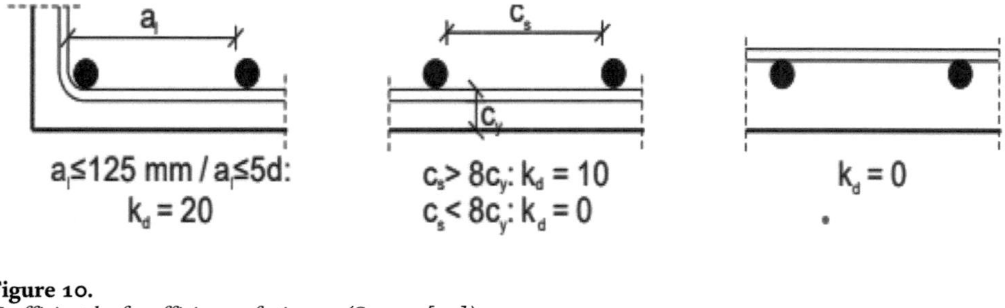

Figure 10.
Coefficient k_d for efficiency of stirrups (Source: [44]).

$$\sum A_{st} = n_g n_t A_{st} \geq \frac{\alpha_t A_{s,cal}}{A_{s,prov}} . n_b A_s \tag{11}$$

Where;
$A_{s,cal}$ calculated area of reinforcement
n_t is the number of stirrups crossing a potential splitting failure surface at a section
$A_{s,prov}$ area of reinforcement provided
n_g number of items of confining reinforcement within the bond length
The design anchorage length l_b can be calculated from Eq. (12):

$$l_b = \frac{\varnothing \sigma_{sd}}{4f_{bd}} \geq l_{b,min} \tag{12}$$

$l_{b,min}$ is the minimum accepted design value for lap length, calculated as:

$$l_{b,min} > max\left\{\frac{0.3\varnothing f_{yd}}{4f_{bd}}; 10\varnothing, 100\ mm\right\} \tag{13}$$

Where σ_{sd} is the stress in the bar that will be anchored by bond across the length of the lap, and is calculated as:

$$\sigma_{sd} = \alpha_1 . f_{yd} \tag{14}$$

$$\alpha_1 = \frac{A_{s,cal}}{A_{s,ef}}$$

Where $A_{s,ef}$ and $A_{s,cal}$ are the actual area of reinforcement and the required area of reinforcement determined in design, f_{yd} is the design yield strength of the reinforcement.

The design lap length is calculated as follows:

$$l_0 = \alpha_4 \frac{\varnothing f_{yd}}{4f_{bd}} \geq l_{0,min} \tag{15}$$

$l_{0,min}$ is the minimum accepted design value for lap length, calculated as:

$$l_{0,min} > max\left\{\frac{0.7\varnothing f_{yd}}{4f_{bd}}; 15\varnothing, 200\ \text{mm}\right\} \tag{16}$$

The coefficient α_4 = 0.7 may be used if no more than 34% of the bars are lapped at the section or the reinforcement stress estimated at the limit state does not surpass 50% of the reinforcement's characteristic strength, otherwise α_4 = 1.0 may be used.

According to Model Code 2010, the stress developed in a lap may be taken as:

$$\sigma_{sd} = l_0/\varnothing.4/\alpha_4.\left[(\alpha_2 + \alpha_3).1.75.(25/\varnothing)^{0.3}.(f_{ck}/25)^{0.5} - 2P_{tr}\right]/\gamma_{cb} \tag{17}$$

Ref [44], specifies two distinct design bond stress-slip relationships in addition to the lap length and design bond strength. The designer determines the suitable relationship to use according to the mode of failure, which is confinement or bond failure. **Table** 7 depicts the general bond stress-slip model.

For a well-confined concrete pull-out failure is expected when the clear spacing between bars is greater than 10Ø and the concrete cover is greater than 5Ø. Another bond-slip model provided in ref. [44] is for splitting failure, the bond strength is obtained using the following expression:

$$\tau_{bu,split} = 6.5\left(f_{cm}/25\right)^{0.25}\left(25/\varnothing\right)^{0.2}\left[\left(c_{min}/\varnothing\right)^{0.33}\left(c_{max}/c_{min}\right)^{0.1} + k_m k_{tr}\right] \quad (18)$$

Equation (4) is used to derived the splitting bond strength given as:

$$\tau_{u,split} = \left(\frac{\varnothing}{l_b}\right)\left(\frac{f_{stm}}{4}\right) = \left(\frac{\varnothing}{l_b}\right)\left(\frac{1}{4}\right)54\left(\frac{f_{cm}}{25}\right)^{0.25}\left(\frac{25}{\varnothing}\right)^{0.2}\left(\frac{l_0}{\varnothing}\right)^{0.55} \quad (19)$$

In addition, ref. [44] provides a coefficient for the influence of cyclic loading, longitudinal cracking, yielding, transverse cracking and stress.

$$\tau_{b,m} = \tau_0 \Omega_y \Omega_{p,tr} \Omega_{cr} \Omega_{cyc} \quad (20)$$

Where
Ω_{cyc} is the effect of cyclic loading
Ω_{cr} is the effect of longitudinal cracking $\Omega_{cr} = 1 - 1.2w_{cr}$
$\Omega_{p,tr}$ is the effect of transverse pressure $\Omega_{p,tr} = 1 - \tanh\left(0.2p_{tr}/0.1f_{cm}\right)$
Ω_y is the effect of yielding

	Splitting		Pull-out
	Stirrups	Unconfined	
τ_{max}	$8(F_{cm}/25)^{0.25}$	$7(F_{cm}/25)^{0.25}$	$2.5\sqrt{f_{cm}}$
s_1	$s(\tau_{max})$	$s(\tau_{max})$	1 mm
s_2	s_1	s_1	2 mm
s_3	$0.5c_{clear}$	$1.2s_1$	C_{clear}
α	0.4	0.4	0.4
τ_f	$0.4\tau_{max}$	0	$0.4\tau_{max}$

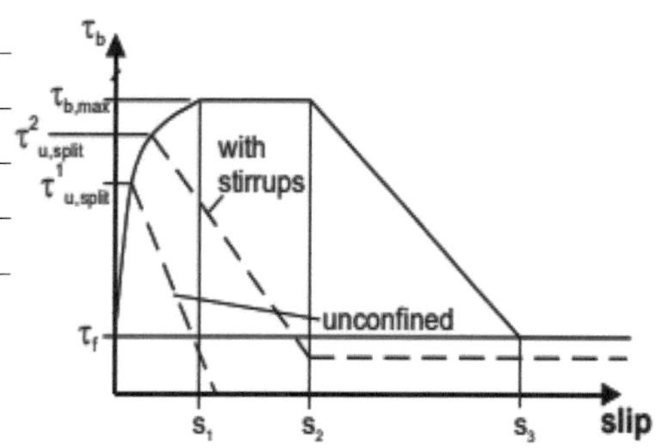

Where:
$\tau_0 = \tau_{max}\left(s/s_1\right)^{\alpha}$ *for* $0 \le s \le s_1$
$\tau_0 = \tau_{max}$ *for* $s_1 \le s \le s_2$
$\tau_0 = \frac{\tau_{bmax}(\tau_{max} - \tau_f)(s - s_2)}{(s_3 - s_2)}$ *for* $s_s \le s \le s_3$
$\tau_0 = \tau_f$ *for* $s_s < s$
c_{clear} *is the clear distance between the ribs*
τ_f *is the residual bond stress*

Table 7.
Bond slip relationship based on [44].

1.3.3 Eurocode 2

The Eurocode 2 design model for laps is based on [51]. In order to determine the design lap length in Eurocode 2, the design bond strength must first be determined, then the anchorage length. The bond strength is used to calculate the anchorage length with the following expression:

$$f_{bd} = 2.25\eta_1\eta_2 f_{ctd} \tag{21}$$

In this equation, η_1 is a coefficient associated with the bar position and bond conditions during concreting, with 0.7 representing all other conditions unity indicating good bond condition. The bond strength to tensile concrete strength ratio is described by the coefficient 2.25. When the diameter of the bar is smaller than 32 mm, the coefficient η_2 is considered as unity; otherwise, the following equation is used:

$$\eta_2 = \frac{132 - \varnothing}{100} \text{ for} \varnothing > 32 \text{ mm} \tag{22}$$

The concrete design strength is f_{ctd} is obtained from $\frac{f_{ctk,o.005}}{\gamma_c}$ where is $f_{ctk,o.005}$ restricted to concrete grade C60/75. The concrete tensile strength is determined as a function of the compressive strength of concrete:

$$f_{ctd} = \frac{f_{ctk,o.005}}{\gamma_c} = \alpha_{ct} = 0.21(f_{ck})^{\frac{2}{3}}/\gamma_c \tag{23}$$

The coefficient α_{ct} which takes into consideration the loading and long-term effect of tensile strength is a nationally defined factor, the value recommended is 1.0. Equation (24) gives the safety considered in the Eurocode 2 anchorage design model. The bond strength is calculated using a 5%- fractile of the concrete tensile strength divided by the concrete's partial safety factor $\gamma_c = 1.5$. Therefore, the average bond strength required for comparing the test results is:

$$f_{bm} = 2.25\eta_1\eta_2 0{,}3 f_{ck}^{\frac{2}{3}} \tag{24}$$

The required basic anchorage length is obtained as:

$$l_{b,rqd} = \left(\frac{\varnothing}{4}\right)\left(\frac{\sigma_{sd}}{f_{bd}}\right) \tag{25}$$

The stress at the cross-section in which the anchorage length begins is the design stress of the bar σ_{sd}. The design length of the anchorage is determined as:

$$l_{bd} = \alpha_1.\alpha_2.\alpha_3.\alpha_4.\alpha_5.l_{b,rqd} \geq l_{b,min} \tag{26}$$

α_1 and α_2 are coefficient associated with bars shape and cover concrete, respectively. $\alpha_1 = 1.0$ for straight rebars.

$$\alpha_2 = 1 - 0.15\left(\frac{(c_{min} - \varnothing)}{\varnothing}\right) \leq 1.0$$

α_3 is the coefficient of shear link if present, with

$$0.7 \leq \alpha_3 = 1 - k\lambda = 1 - \frac{k(\sum A_{st} - \sum A_{st,min})}{A_s} \leq 1.0 \tag{27}$$

If there is no transverse pressure and welded transverse reinforcement, α_4 and α_5 can be taken as unity. However, when transverse pressure is present, the anchorage length can be decreased by:

$$0.7 \leq \alpha_5 = 1 - 0.004p \leq 1.0. = 1 - 0. \tag{28}$$

K accounts for the transverse reinforcement's efficacy in relation to its position inside the section. The difference in cross-section area between the minimum transverse reinforcement $\sum A_{\text{st, min}}$ and the transverse reinforcement provided along the anchorage length is described by the coefficient λ with $\sum A_{\text{st}} = 0.25A_{\text{s}}$ for anchorage and $\sum A_{\text{st, min}} = 1.0A_{\text{s}}\left(\frac{\sigma_{\text{sd}}}{f_{\text{yd}}}\right) \geq 1.0A_{\text{s}}$ for laps.

Eurocode 2 recommended that laps should be positioned in a low moment region and staggered. The clear lapped spacing between bars should not exceed 50 mm or 4∅. If all bars are in the layer, the permissible proportion of lapped bars in tension is 100% and should not exceed 50% for lapped bars in several layers.

The basic required anchorage length and the coefficients are included in the design of lap length $l_{\text{b,rqd}}$, which takes into account the key impacting parameters.

$$l_0 = \alpha_1.\alpha_2.\alpha_3.\alpha_4.\alpha_5.\alpha_6 l_{b,rqd} \geq l_{0,min} \tag{29}$$

The coefficient α_1 to α_5 described above. For the proportion of bars lapped at a section, the coefficient α_6 can be taken from **Figure 11**.

The coefficient α_6 may also be determined as follows:

$$1.0 \geq \alpha_6 = \left(\frac{\rho_1}{25}\right)^{0.5} \leq 1.5 \tag{30}$$

		Bond condition, (see Figure 1)	Reinforcement in tension, bar diameter, ϕ (mm)								Reinforcement in compression
			8	10	12	16	20	25	32	40	
Anchorage length, l_{bd}	Straight bars only	Good	230	320	410	600	780	1010	1300	1760	40ϕ
		Poor	330	450	580	850	1120	1450	1850	2510	58ϕ
	Other bars	Good	320	410	490	650	810	1010	1300	1760	40ϕ
		Poor	460	580	700	930	1160	1450	1850	2510	58ϕ
Lap length, l_0	50% lapped in one location (α_6 = 1.4)	Good	320	440	570	830	1090	1420	1810	2460	57ϕ
		Poor	460	630	820	1190	1560	2020	2590	3520	81ϕ
	100% lapped in one location (α_6 = 1.5)	Good	340	470	610	890	1170	1520	1940	2640	61ϕ
		Poor	490	680	870	1270	1670	2170	2770	3770	87ϕ

Notes

1 Nominal cover to all sides ≥ 25 mm, [i.e. $a_2 \leq 1$]. At laps, clear distance between bars ≤ 50mm.

2 $\alpha_1 = \alpha_3 = \alpha_4 = \alpha_5 = 1.0$. For the beneficial effects of shape of bar, cover and confinement see Eurocode 2, Table 8.2.

3 Design stress has been taken as 435 MPa. Where the design stress in the bar at the position from where the anchorage is measured, σ_{sd}, is less than 435 MPa the figures in this table can be factored by σ_{sd}/435. The minimum lap length is given in cl 8.7.3 of Eurocode 2.

4 The anchorage and lap lengths have been rounded up to the nearest 10 mm.

5 Where 33% of bars are lapped in one location, decrease the lap lengths for '50% lapped in one location' by a factor of 0.82.

6 The figures in this table have been prepared for concrete class C25/30; refer to Table 13 for other classes or use the following factors for other concrete classes.

Concrete class	C20/25	C28/35	C30/37	C32/40	C35/45	C40/50	C45/55	C50/60
Factor	1.16	0.93	0.89	0.85	0.80	0.73	0.68	0.63

Figure 11.
Lap and anchorage length for concrete class C25/30 (mm). (Extract from [45]).

with

ρ_1 is the proportion of lapped bars at a section

The placing of shear links at the outer section of the lap length is required by Eurocode 2 for the concentration of splitting forces at lap ends. Shear links required for other reasons might be assumed sufficient if the proportion of lapped bars is less than 25%. The cross-sectional area of the shear link must not be lower than the cross-sectional area of one lapped bar for laps with a diameter higher than or equal to 20∅.

Equation (29) must be rearranged to obtain the bar developable stress. This enables the computation of experimental data and compared the various design model.

$$\sigma_{sd} = (l_0/\varnothing)\left(\frac{4f_{bd}}{\alpha_1\alpha_2\alpha_3\alpha_3\alpha_5\alpha_6}\right) \tag{31}$$

For straight bars without shear pressure and good bond condition, the average bar stress may be calculated as:

$$\sigma_{sd} = (l_0/\varnothing)\left(\frac{2.7f_{ck}^{2/3}\eta_2}{\alpha_2\left(1-k\left(\frac{\sum A_{st}-\sum A_{st,min}}{A_s}\right)\right)\alpha_6}\right) \tag{32}$$

1.3.4 American concrete institute

Reference [52] developed the design model that is used in American Concrete Institute (ACI). The design anchorage length is determined as:

$$\sigma_{sd} = \left(\frac{3}{40}\right)\left(f_y/\lambda\sqrt{f_c'}\right)\left(\frac{\psi_t\psi_c\psi_s}{\frac{c_b+k_{tr}}{\varnothing}}\right)\varnothing \tag{33}$$

f_y is the bar yield stress

$c_b = \min\left\{\frac{(c_s+\varnothing)}{2}; c_y + \frac{\varnothing}{2}; c_x + \frac{\varnothing}{2}\right\}$, because the cover values are connected to the bar centre in ACI, the values $\frac{\varnothing}{2}$ have to be added to comply with the notion.

f_c is the cylinder concrete strength limited to 69 MPa

ψ_c is the coefficient for coated reinforcement (for uncoated bars $\psi_c = 1.0$)

ψ_s is the coefficient for bar diameter (1.0 for bars ≥ 22 mm, otherwise 0.8)

ψ_t is the coefficient for bond (for good bond condition $\psi_t = 1.0$)

k_{tr} is shear link index

$$k_{tr} = \frac{40A_{st}[in.^2]}{s[in.]n_b} \tag{34}$$

$$\frac{c_b + k_{tr}}{\varnothing} \tag{35}$$

Where

S is the shear link spacing

A_{st} is the cross-sectional area of all shear links within the spacing

n_b are the number of lapped bars

Pull-out failure becomes more probable for confinement ratios $\frac{(c_b+k_{tr})}{\varnothing}$ above 2.5, and increasing transverse reinforcement or concrete cover does not result in enhanced bond capacity. [53] reported that for $\frac{c_b+k_{tr}}{\varnothing} > 3.75$, the bond capacity did not increase, thus the limit of 2.5 provides extra safety [50].

For normal weight concrete, uncoated reinforcement, and good bond conditions, Eq. (33) can be simplified to

$$l_b = \left(\frac{3}{40}\right)\left(\frac{f_y}{\sqrt{f'_c}}\right)\left(\frac{\psi_s}{\frac{c_b+k_{tr}}{\varnothing}}\right)\varnothing(in.) \tag{36}$$

Rearranging for the anchorage strength gives

$$\sigma_s = \left(\frac{40}{4}\right)\left(\frac{l_b}{\varnothing}\right)\sqrt{f'_c}\left(\frac{c_b+k_{tr}}{\varnothing}\right)\left(\frac{1}{\psi_s}\right)(\text{psi}) \tag{37}$$

If confinement by a compressive reaction is present at the simple support, the development length can be decreased by around 30%. It is worth noting that the ACI design guide permit bar diameters of up to 57 mm. For reinforcing bars with 43 and 57 mm diameters, the minimum allowable cover concrete is 38 mm in beams and 19 mm in slabs.

It must be noted that ref. [50] allows for bar diameters up to 57 mm. The minimum permissible concrete cover is 38 mm in beams and 19 mm in slabs (for Ø 43 mm and Ø 57 mm reinforcing bars: 38 mm). The smaller of the two values, the bar diameter or 25 mm, is the minimum clear bar spacing.

Due to a lack of sufficient experimental evidence, ACI does not allow laps of ACI does not permit laps of 43Ø and 57Ø. **Table 8** shows the relation between the required and provided reinforcement and the coefficient for the proportion of bars lapped based on ACI recommendation. The coefficient 1.3 is not established on bond stress investigations, however, it is intended to promote the placement of laps away from high tensile stress regions to places where the reinforcement cross-sectional area provided as a minimum is twice that required by analysis. As a result, this coefficient contains a level of safety.

Smaller bar diameters with short lap lengths, the majority of which were less than 300 mm, were used to validate the design model. As a result, a factor for bar size γ is 1.0 and 0.8 for larger bar diameters and smaller bars not more than 22 mm were introduced. The ACI committee 408 advices against a size effect factor of less than 1.0.

1.3.5 German national annex

The coefficient for the minimum permissible bar spacing c_s and the tensile strength α_{ct} are the nationally determined parameters for lap and anchorage calculation for

Lap length	$A_{s,prov}/A_{s,req}$	Percentage of A_s spliced
max $\{1.0.\ l_b; 305\ \text{mm}\}$	≥ 2.0	50
max $\{1.3.\ l_b; 305\ \text{mm}\}$		100
	< 2.0	All cases

Table 8.
Spliced length coefficient for the percentage of lapped bars and the provided reinforcement [50].

Eurocode 2. The National annex defines the bars spacing as $c_s = 1.\varnothing$, and the tensile strength $\alpha_{ct} = 1.0$ for bond strength calculation.

The anchorage length at direct supports may be computed using $\alpha_5 = 2/3$ taking transverse pressure into consideration. For anchorages, National annex suggests using a simple cover concrete coefficient of $\alpha_2 = 1.0$. The cover concrete orthogonal to the lap plane c_y is not taken into consideration for spliced rebars with straight ends, based on the National annex. For laps, $c_{\min} = \min\left\{\frac{c_s}{2}0; c_x\right\}$ is the cover used for the calculation of the coefficient α_2. The lap factor α_6 is determined by the bar diameter and the proportion of bars lapped at a location. The recommended values for α_6 are shown in **Table 9**.

If more than 50% of the reinforcement is spliced at one location, according to Eurocode 2, the transverse reinforcement in laps shall be made by links. The National annex relaxes this requirement by specifying that the transverse reinforcement does not need to have a longitudinal spacing of $0.5l_0$ between the adjacent laps centres or consist of links with the distance between adjacent laps greater than $10\varnothing$.

1.3.6 PTI working draft

The PTI working draft offers a revised design model based on Fib Bulletin 72 for the next generation of Eurocode 2. For ease of use, the exponents were simplified. The required bond strength for good bond conditions comes from

$$l_{bd,req.} = 40\left(25\,MPa/f_{ck}\right)^{1/2}\left(\frac{\sigma_{sd}}{435MPa}\frac{\gamma_c}{1.5}\right)^{3/2}\left(\varnothing/20\,mm\right)^{1/3}\left(1.5\varnothing/c_{d,c\,onf}\right)^{1/2} \le l_{b,min} \tag{38}$$

with

$l_{\mathrm{b,min}}$ is the minimum lap length with $20\varnothing$ for laps

$$c_{d,c\,onf} = c_d + \left(30k_{conf}\frac{n_l A_{st}}{n_b \varnothing s_{st}} + 8\sigma_{ctd}/\sqrt{f_{ck}}\right)\varnothing \le 3.75\varnothing \tag{39}$$

σ_{ctd} is the design value of the mean compression stress perpendicular to the potential splitting plane

K_{conf} is the effectiveness factor. K_{conf} is taken as 0.25 for shear links within the cover c_y with cover spacing greater than $8\varnothing$, and 1.0 for confinement reinforcement crossing the potential splitting plane (the maximum recommended distance from the leg to the lapped bar is less than $5\varnothing$). In other circumstances, K_{conf} is taken as zero.

Percentage of bars lapped at a section in one layer		**Bar diameter ∅**
≥ 33%	≤ 33%	
1.4 (1.0)	1.2 (1.0)	< 16 mm
2.0 (1.4)	1.4 (1.0)	≥ 16 mm

The values in brackets are valid for $c_s \geq 8\varnothing$ and $c_x \geq 4\varnothing$.

Table 9.
German National Annex for the percentage of a spliced bar under tension based on [45].

The PTI working draft recommended the laps be designed for $1.2 \times \sigma_{sd}$ if tension laps are located in regions where the yield strength may be exceeded. Therefore, a reduction of lapped bars or confining reinforcement is required. Reliability analysis of the coefficients 8,30, and 40 in expressions (38) and (39). Rearranging for the developable stress in anchorages gives

$$f_{std} = 435(1.5/\gamma_c)^{1/3}(20/\varnothing)^{2/9}(l_b/40\varnothing)^{2/3}\left(c_{d,conf}/1.5\varnothing\right)^{1/3} \quad (40)$$

1.3.7 Canbay and Frosch

Four hundred and eighty experiments with and with shear links were used by [54] to verify their model for ultimate lap load. The ultimate lap strength at splitting failure, taking shear link into account is defined as

$$\sigma_{sd} = F_{split} + F_{st}/n_b A_s \tan\beta = 2.75/n_b A_s\left(F_{split} + F_{st}\right)(\text{ksi}) \quad (41)$$

Where

θ inclination of struts commencing at the rib flanks (20 degrees provided optimal results)

F_{st} is the splitting resistance by a shear link [kip]

F_{split} is the splitting resistance by cover concrete along the lap length [kip]

[54] distinguish side (**Figure 12**, middle) and face splitting types (**Figure 12**, left).

It is assumed that the tensile concrete stress surrounding the bars is linear along the lap length. The bond strength nonlinearity along the splice length is taken into account by using an efficient lap length l_0^*. The side-splitting force $F_{split,side}$ is

$$F_{split,side} = l_0^*\left[2c_x^* + 2c_x^*(n_b - 1)\right]6f_c^{1/2}(\text{kip}) \quad (42)$$

For face-splitting, the splitting force $F_{split,face}$ is

$$F_{split,face} = l_0^*\left[2c_y^*\left(0.1\frac{c_x}{c_y} + 0.9\right) + 2c_y^*(n_b - 1)\left(0.1\frac{c_s}{2c_y} + 0.9\right)\right]6f_c^{1/2} \quad (43)$$

With

$c_y^*, c_x^*, (\frac{c_s}{2})^*$ coefficients for the efficiency of concrete cover at linear stress distribution

$$l_0^* = l_0\frac{9.5I}{\sqrt{l_0/\varnothing f_c^{\frac{1}{4}}}} \le l_0$$

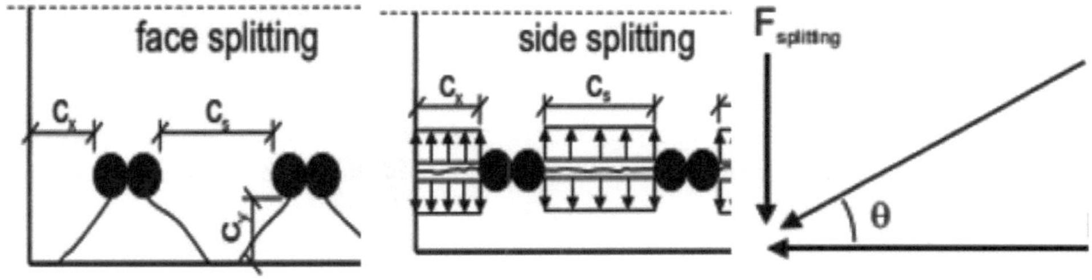

Figure 12.
Face splitting (left), side splitting (middle) and force distribution at bond forces (right). [Adapted from [54]].

l_0 coefficient for the efficiency of lap length at linear stress distribution

In the failure of splitting, shear links provide extra resistance. The splitting force $F_{st,\ side}$ provided by the shear link, is given by expression (44). For side splitting, this expression includes the number of legs crossing the splitting plane in

$$F_{st,side} = n_{st} n_l A_{st} f_{yt} \left[\frac{n_b f_c^{1/2}}{170} \right] (\text{kip}) \tag{44}$$

Because the shear link crosses the splitting plane at each bar in face split failure, expression (45) incorporates the bar number instead of the leg number in determining the splitting force.

$$F_{st,face} = n_{st} n_b A_{st} f_{yt} \left[\frac{n_b f_c^{1/2}}{170} \right] (\text{kip}) \tag{45}$$

62 MPa is the suggested transverse bar yield strength. A safety factor of 1.2 was utilised to develop the design expression (41). 50% of the calculated strength was unsafe without the safety factor. Only 16% of the confined and 10% of the unconfined test meet their yield strength when the factor of 1.2 was used. This take into consideration the lap length designed based on the proposed model and nominal yield strength of 414 MPa.

2. Conclusions

This chapter provides the context and presents the state of art for using stainless steel as a structural material. It is demonstrated that stainless steel is a remarkable building material that is becoming a more desirable option for RC structures because of its recyclability, long life cycle favourable mechanical properties, ductility and excellent corrosion resistance. There are numerous factors preventing stainless steel reinforcement from being used more frequently in reinforced concrete structures. Firstly, mainly due to this is a new topic in structural engineering, there is a lack of performance data and design guidance available in the public domain. The second issue is that engineers have the perception that stainless steel reinforcement is expensive. Despite having a higher initial cost than carbon steel, stainless provides efficient and a very competitive design option throughout the duration of a structure's lifetime when rehabilitation and maintenance expenses are taken into account.

It is important that structurally effective design solutions are available that take into account and exploit the advantageous and distinctive properties of stainless given its high initial cost. Unfortunately, current design guidance, such as Eurocode 2, does not have an effective method for designing structures with stainless steel reinforcement; rather, they include inadequate material models for reinforcing bar that do not make use of the unique properties of stainless steel. While this premise could be valid for concrete with carbon steel reinforcement, it provides widely erroneous estimates when suing stainless steel reinforcement. This is mostly due to stainless steel's early stage of nonlinear behaviour and its high strain hardening. The design of stainless steel reinforced concrete members using the current design criteria for carbon steel RC structures is thus neither accurate nor efficient.

Additionally, it has been reported that there is little discussion of the bond behaviour of stainless-steel reinforcement in the literature and that the research that is available is both conflicting and limited. The existing design guidelines, such as Model Code 2010 and Eurocode 2, generally recommend employing the same standard as for traditional carbon steel reinforcement and do not provide particular lap length design guidelines for designing stainless steel RC structure without appropriate test data may not be a safe option because it has been observed that stainless steel reinforcement may develop a lower bond strength than carbon steel reinforcement.

Conflict of interest

The authors declare no conflict of interest.

Author details

Yakubu Mustapha Karkarna[1], Ali Bahadori-Jahromi[1*], Hamid Zolghadr Jahromi[2], Emily Halliwell[3] and Musab Mohammad Rabi[4]

1 School of Computing and Engineering, University of West London, London, United Kingdom

2 School of Computing and Engineering, University of Westminster, London, United Kingdom

3 The Concrete Centre Structural Engineer, London, United Kingdom

4 Department of Civil Engineering, Jerash University, Jerash, Jordan

*Address all correspondence to: ali.jahromi@uwl.ac.uk

References

[1] Shamass R, Cashell K. Analysis of stainless steel-concrete composite beams. Journal of Constructional Steel Research. 2018;**152**:132142

[2] Baddoo NR. Stainless steel in construction: a review of research, applications, challenges and opportunities. Journal of Constructional Steel Research. 2008;**64**(11):1199-1206

[3] Gardner L. The use of stainless steel in structures. Progress in Structural Engineering and Materials. 2005;7(2): 45-55

[4] Eladly MM. Behaviour of stainless steel beam-to-column bolted connections–Part 1: simplified FE model. Journal of Constructional Steel Research. 2020;**164**:105-784

[5] Rabi M, Cashell KA, Shamass RJES. Flexural analysis and design of stainless steel reinforced concrete beams. Engineering Structures. 2019;**198**:109432

[6] EN 10088-2. Stainless steels—Part 2: technical delivery conditions for sheet/ plate and strip of corrosion resisting steels for general purposes. European Committee for Standardization (CEN). 2014

[7] Evans K. RB rebak in corrosion science—A retrospective and current status in honor of Robert P. Corrosion Science. 2002;**13**:344-354

[8] Ramberg W, Osgood WR. Description of Stress-Strain Curves by Three Parameters. Technical Note No. 902, National Advisory Committee for Aeronautics, Washington DC. 1943

[9] Rabi M, Cashell KA, Shamass R. Ultimate behaviour and serviceability analysis of stainless steel reinforced concrete beams. Engineering Structures. 2021;**248**:113259

[10] Markeset G, Rostam S, Klinghoffer O. Guide for the use of stainless steel reinforcement in concrete structures. Norway. 2006

[11] Sharif AM, Al-Mekhlafi GM, Al-Osta MA. Behavior of circular stainless steel stub columns internally strengthened by longitudinal carbon steel bars. Engineering Structures. 2019;**199**: 109-617

[12] Rabi M, Cashell KA, Shamass R. Analysis of concrete beams reinforced with stainless steel. In: Proceedings of the Fib Symposium 2019: Concrete-Innovations in Materials, Design and Structures. 2019. pp. 690-697

[13] Nationwide Stainless. 2021. Stainless Steel Rebar. Available from: https://www.nationwidestainless.co.uk/products/stainless_steel_rebar/

[14] Rabi MM. Analysis and Design of Stainless Steel Reinforced Concrete Structural Elements [doctoral dissertation]. London: Brunel University; 2020

[15] Baddoo R, Burgan A. Structural Design of Stainless Steel SCI, Publication No. P291. 2012

[16] Rabi M, Cashell KA, Shamass R, Desnerck P. Bond behaviour of austenitic stainless steel reinforced concrete. Engineering Structures. 2020; **221**:111027

[17] Rabi M, Shamass R, Cashell KA. Structural performance of stainless steel reinforced concrete members: A review. Construction and Building Materials. 2022;**325**:126673

[18] Cramer S, Covino B, Bullard S, Holcomb G, Russell J, Nelson F, et al. Corrosion prevention and remediation strategies for reinforced concrete coastal bridges. Cement and Concrete Composites. 2002;**24**(1):101-117

[19] Rabi M, Shamass R, Cashell KA. Experimental investigation on the flexural behaviour of stainless steel reinforced concrete beams. Structure and Infrastructure Engineering. 2022: 1-13

[20] British Stainless Steel Association. Classification of stainless steel types. 2000. Available from: https://www.bssa.org.uk/cms/File/SSAS1.2-Classification%20of%20Stainless%20Steels.pdf

[21] Aalco. Specifications for Stainless Steel Products. 2013

[22] Metals4U. 2021. Stainless steel reinforcing bar. Available from: https://www.metals4u.co.uk/stainless-steel/c8/reinforcing-bar/c1875

[23] Gardner L, Bu Y, Francis P, Baddoo NR, Cashell KA, McCann F. Elevated temperature material properties of stainless steel reinforcing bar. Construction and Building Materials. 2016;**114**:977-997

[24] Gedge G. Rationale for using stainless steel reinforcement in the UK construction industry. UK. 2003

[25] McGurn J. Stainless steel reinforcing bars in concrete. In: Proceedings of the International Conference of Corrosion and Rehabilitation of Reinforced Concrete Structures. Orlando. 1998

[26] British Highways Authority. Advice note: design manual for roads and bridges BA84/02: the use of stainless steel reinforcement in highway structures. 2003. Available from: https://www.bssa.org.uk/cms/File/REBar%20report.pdf

[27] Fan S, Ding X, Sun W, Zhang L, Liu M. Experimental investigation on fire resistance of stainless steel columns with square hollow section. Thin-Walled Structures. 2016;**98**:196-211

[28] Castro H, Rodriguez C, Belzunce F, Canteli A. Mechanical properties and corrosion behaviour of stainless steel reinforcing bars. Journal of Materials Processing Technology. 2003;**143**: 134-137

[29] Medina E, Medina JM, Cobo A, Bastidas DM. Evaluation of mechanical and structural behavior of austenitic and duplex stainless steel reinforcements. Construction and Building Materials. 2015;**78**:1-7

[30] DMRB. DMRB 1.3.12. Design manual for roads and bridges (DMRB). 2003

[31] BA 84/02. Design manual for roads and bridges part 15: use of stainless steel reinforcement in highway structures. Highways Agency UK. 2003

[32] Bautista A, Blanco G, Velasco F, Gutiérrez A, Palacín S, Soriano L, et al. Passivation of duplex stainless steel in solutions simulating chloride contaminated concrete. Materiales de Construcción. 2007;**57**(288):17-32

[33] Pérez-Quiroz J, Terán J, Martínez M, Genescá J. Assessment of stainless steel reinforcement for concrete structures rehabilitation. Journal of Constructional Steel Research. 2008;**64**(11):1317-1324

[34] Huang Y, Young B. Stress–strain relationship of cold-formed lean duplex stainless steel at elevated temperatures. Constructional Steel Research. 2014;**92**: 103-113

[35] Lopes N, Real PV, da Silva LS, Franssen J. Numerical analysis of stainless steel beam-columns in case of fire. Fire Safety Journal. 2012;**50**: 35-50

[36] Tondini N, Rossi B, Franssen J. Experimental investigation on ferritic stainless steel columns in fire. Fire Safety Journal. 2013;**62**:238-248

[37] Helland S. Design for service life: implementation of fib Model Code 2010 rules in the operational code ISO 16204. Strcural Concrete. 2013;**14**(1):10-18

[38] García-Alonso M, Escudero M, Miranda J, Vega MI, Capilla F, Correia M, et al. Corrosion behaviour of new stainless steels reinforcing bars embedded in concrete. Cement and Concrete Research. 2007;**37**(10): 1463-1471

[39] Nurnberger U. Stainless Steel in Concrete: State of the Art Report. 1996

[40] Serdar M, Žulj LV, Bjegović D. Long-term corrosion behaviour of stainless reinforcing steel in mortar exposed to chloride environment. Corrosion Science. 2013;**69**:149-157

[41] Bond AJ, Brooker O, Harris AJ, Harrison T, Moss RM, Narayanan RS, et al. How to Design Concrete Structures to Eurocode 2. 2nd ed. London: MPA The Concrete Centre; 2018

[42] Zhou Y, Ou Y, Lee GC. Bond-slip responses of stainless reinforcing bars in grouted ducts. Engineering Structures. 2017;**141**:651-665

[43] Ahlborn T, DenHartigh T. Comparative bond study of stainless and highchromium reinforcing bars in concrete. Transportation Research Record: Journal of the Transportation Research Board. 2003;**1845**:88-95

[44] Model Code for Concrete. 2013. Fib Model Code for Concrete Structures 2010. fib Model Code for Concrete Structures 2010. Wiley. DOI: 10.1002/9783433604090

[45] EN 1992-1-1. Eurocode 2: design of concrete structures part 1-1: General rules and rules for buildings. European Committee for Standardization (CEN). 2004

[46] PT.SC2.T1. PTI working draft prEN 1992-1-1:2018. European Committee for Standardization. European Committee for Standardization; CEN, European Committee for Standardization. 2018

[47] Fib Bulletin 72. Bond and anchorage of embedded reinforcement: Background to the fib Model Code for Concrete Structures 2010. Lausanne, Switzerland. 2014

[48] Eligehausen R. Übergreifungsstöße zugbeanspruchter Rippenstäbe mit geraden Stabenden. Stuttgart Germany: Wilhelm Ernst & Sohn; 1979

[49] Ferguson PM, Briceno EA. Tensile lap splices part 1: Retaining wall type varying moment zone. No. 113-2. Centre for Highway Research. Texas. 1969

[50] ACI committee 318. Building Code Requirements for Structural Concrete and Commentary (ACI 318M-11). American Concrete Institute, Farmington Hills, MI. 2011

[51] CIB-FIP Model Code. CEB-FIP model code 1990: design code [online]. 1991. Bulletin d'Information. London: Thomas Telford Services Ltd. Available from: http://books.google.com/books?hl=en&lr=&id=IRG9GTPJ7s8C&oi=fnd&pg=PR5&dq=CEB-FIP+model+code+1990:+Design+Code&ots=756sbOWzgo&sig=x2_W4HCZLIqwahl62aMrKidAcTU [Accessed: July 22, 2014]

[52] Orangun CO, Jirsa JO, Breen JE. A reevaulation of test data on development length and splices. ACI Journal Proceedings. 1977;**74**(3):114-122

[53] Darwin D et al. Development length criteria for conventional and high relative rib area reinforcing bars. ACI Structural Journal. 1996;**93**(3):512-516

[54] Canbay E, Frosch RJ. Bond strength of lap-spliced bars. ACI Structural Journal. 2005;**102**(4):605-614. DOI: 10.14359/14565

Section 3

Pathologies and Identification of Reinforced Concrete

Chapter 12

Study of the Application of Terrestrial Laser Scanning for Identification of Pathologies in Concrete Structures

Sérgio Orlando Antoun Netto,
Lucas Pires Chagas Ferreira de Carvalho,
Ana Waldila de Queiroz Ramiro Reis,
Leonardo Vieira Barbalho and Lucas de Campos Rodrigues

Abstract

Laser scanning enhances classic field surveys. It is a remote, active, noninvasive, nondestructive, and high-precision technique to capture reality that records from thousands to millions of points per second in a detailed representation of the situation called a point cloud. The surveys are performed along the object of interest in a process called scanning, which has as its gross product a dense cloud of three-dimensional points of the scanned object. This point cloud stores information about the object's geometry, return pulse intensity, and point color data. As a way of extending the uses of terrestrial laser scanning, this work studies the application of this method in civil engineering, through the identification of pathologies in reinforced concrete structures, aiming to show how geoinformation can be employed in this area. To this end, a case study of the São Cristóvão Viaduct was conducted in the city of Rio de Janeiro. This study included a definition of the site of analysis; planning and execution of the field survey to collect raw data; processing of the point cloud; and generation of a three-dimensional surface for global visualization of the structure and identification of pathological manifestations and the regions where they were observed.

Keywords: terrestrial laser scanning, point clouds, concrete structures, pathology of structures, remote sensing

1. Introduction

The technology using laser scanners is an improvement for field surveys. In classical topography survey, using angles and displacements in three-dimensional surveys, the final objective is to determine coordinate points. Initially, terrestrial surveys used theodolites and the technique was efficient when introducing electronic measurements of displacement and directions. The integration between both techniques led to total stations, employing reflective prisms to determine three-dimensional coordinate points on terrestrial surfaces. Subsequently, the evolution of laser measurement allowed finding three-dimensional coordinate points without prisms. These points are calculated remotely, which is useful mainly in places with difficult access or risk to the operator's safety. These factors led to the development of robotic total stations, where coordinates can be measured without operators [1].

In this context, terrestrial laser scanning (TLS) was developed for remote sensing to measure displacements and determine coordinate points without prisms or contact with the object of interest. These measurements are not made at a specific point, but over the entire object, in a process called scanning. The results are three-dimensional "point clouds" (x, y, z) of the object. Besides that, the devices can register the pulse's intensity when it returns to the lens, and those with digital cameras integrated can register the color of points, which results in spectral information for each point [1, 2].

This characteristic makes TLS a versatile technique, with various applications, such as engineering, architecture, and geology. Besides that, the equipment has high data acquisition rates (thousands or millions of points per second), which can provide high agility at construction sites. This is a noninvasive and nondestructive technique, where there is no contact between the equipment and the object of analysis [2].

As a way of extending the use of terrestrial laser scanning, this work presents an application of TLS in civil engineering, specifically to identify pathologies in reinforced concrete structures. For this, we performed a study of the São Cristóvão Viaduct, in the north zone of the city of Rio de Janeiro.

The data gathered were transformed into a "point cloud," for visualization of the structure and identification of pathologies.

Reinforced concrete structures in general develop pathologies due to external factors, such as weather and human actions. In Brazil, analysis of the structural integrity of concrete structures is deficient, leading to precarious conservation due to the failure to detect pathologies and avoiding disasters such as building collapses. In this context, the use of TLS for semi-automatic detection and classification of pathologies can be a relevant advance to prevent disasters.

So, the general objective of this work is to study the applicability of terrestrial laser scanning for identification of pathologies in reinforced concrete structures, to demonstrate how remote sensing can be used in civil engineering in complementation with visual inspection to improve detection of pathologies and thus conservation of structures.

This chapter is divided into five more sections. The second presents the theoretical grounds of terrestrial laser scanning, with a brief history of remote sensing and information about the devices, their manufacturers, and specifications, concluding with the main applications presented in the literature.

Section 3 provides theoretical aspects of reinforced concrete structures, a brief history of concrete materials and their composition, and main characteristics and their structural behavior. This is complemented by the main characteristics of steel and the relevant Brazilian standards, followed by introduction of the degradation process and pathologies found in concrete structures.

Section 4 presents a case study of São Cristóvão Viaduct in the city of Rio de Janeiro city. This includes the description of the main structural characteristics and the equipment employed, the data obtained, and their processing.

Section 5 presents the results obtained from processing the data gathered by the terrestrial laser scanner to identify pathologies.

Finally, the work concludes with discussion about laser scanning's future applicability and suggestions for future research.

2. Terrestrial laser scanning

2.1 Remote sensing

There are several ways to conceptualize remote sensing. This can be defined as a technique that uses sensors to capture and record at a distance, without direct contact between the sensor and the object, the energy reflected or absorbed by features of interest on the Earth's surface [3]. Florenzano [4] defined remote sensing as the technology that allows acquisition of images and other types of data from a given surface by capturing and recording the energy reflected or emitted by it. This energy is electromagnetic radiation (REM), measured in frequency and wavelength.

In this work, the studied remote sensing technique is known as LIDAR (Light Detection and Ranging). This technique is similar to radar (Radio Detection and Ranging), the difference being that it uses laser beams instead of radio waves to detect and map objects [5]. LIDAR is a technology that uses active remote sensing equipment, which emits electromagnetic radiation to the object of interest and measures the radiation that is reflected to a receiver after interacting with the Earth's atmosphere or objects [6]. This technology has been perfected from the fundamentals of remote sensing [7].

LIDAR sensors can operate in the visible and middle infrared ranges of the electromagnetic spectrum. For cartographic purposes, the technology can be used from the ground or airborne platforms [8]. According to Longley et al. [6], terrestrial laser scanners can also operate in the ultraviolet range.

2.2 Fundamentals of terrestrial laser scanning

As mentioned, laser scanning is a remote and active technique of high-precision reality capture that records multiple data simultaneously, in a detailed representation of the object of interest, called a point cloud. It is considered a remote method because there is no contact between the operator and the object of interest. The ability of the scanner to generate its own electromagnetic radiation makes it an active system. According to Bordin [9], the ground laser scanner is based on the stimulated emission of radiation, and its light beam is of the laser type.

Laser devices (from Light Amplification by Stimulated Emission of Radiation) were invented in 1960 by Theodore H. Maiman, from two events in the twentieth century. According to Brandalize [10], Albert Einstein, in an article published in 1917, suggested the stimulated emission of light radiation (atoms of light sources) for the emission of photons, which made possible the creation of amplifiers and oscillators. In 1954, MASER (Microwave Amplification by Stimulated Emission of Radiation) was invented as a result of research with radar during World War II. These events made possible the appearance, among other devices, of the laser scanner, with help from the development of computing.

Laser scanning has been applied in several areas of knowledge, especially in engineering, through geodetic and topographic sciences, among others [7]. In general, this term describes any type of technology that measures distances repeatedly and accurately (through time measurement). These measurements are grouped in a point cloud based on the calculated coordinates of each point.

The laser scanner is a remote sensing device used to measure the distance between the sensor and the surface of objects. Its operation is based on the use of laser beams that are emitted toward the targets ([11] cited in [7]). These beams are directed to the environment with the aid of a mirror, or prism, and a motor that generates the horizontal rotation around the vertical axis of the instrument [12], as shown in **Figure 1**.

In addition to the vertical movement generated by mirrors or prisms, controlled horizontal movement is also required, usually implemented through the use of a motor, allowing a series of profiling around the vertical axis of the instrument [12], as illustrated in **Figure 1**.

During a scan, each laser pulse projects a circle (or an ellipse, depending on the beam's angle of incidence) onto the target, called the instantaneous coverage area. The diameter of this circle, or the average of the minor and major axis of the ellipse, depends on the divergence of the emitted pulse. The divergence is the physical characteristic of the laser diffusion as it spreads. In the case of the laser, this diffusion is smaller than the case of a light beam or radio wave, so that it is a projection on the target has small dimensions, even at high distances between the sensor/emitter and the target, and can be adjusted by means of optical elements in the equipment. The return signal depends on the energy dispersion of the incident pulse within the area where the pulse projects onto the target [10].

The beams that reach the object of interest interact with their surfaces. This interaction varies according to the features of the object. Different features generate different returns, according to the shape, dimensions, and constituent material. In this way, each pulse can generate a return or multiple returns [8].

At the end of a scan, for each point collected, the laser scanner registers three-dimensional (X, Y, Z) coordinates, oriented to the phase center of the device (horizontal and vertical axis intersection). In addition, the return signal strength

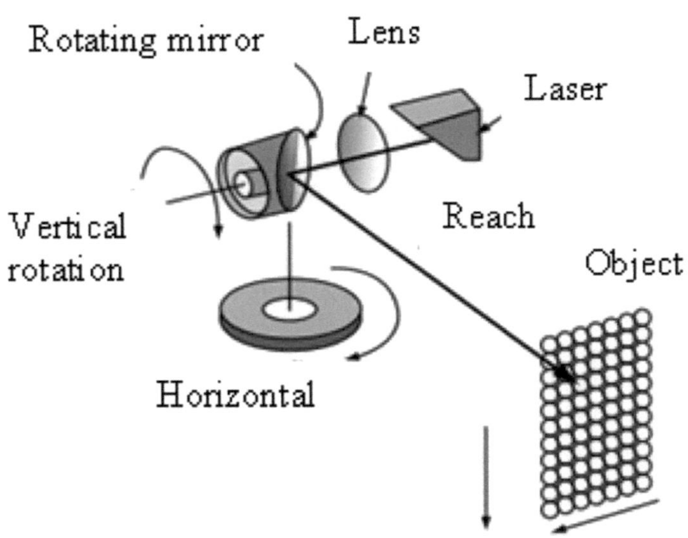

Figure 1.
Projection of laser beams to the object, from the mirror (adapted from [12]).

of each point on the surface scanned by the laser is also computed. This return pulse intensity indicates voltage peaks of the signals that are reflected by the objects, recorded by the equipment. In this way, the intensity is a measurement of the number of photons (energy) that is captured by the sensor. It depends on the type of material that composes the object and is affected by three parameters: the distance to the object, its reflectance, and the angle of incidence of the emitted pulse. Therefore, the quality of the reflected pulses depends directly on the range quality of the emitted pulses [8].

From the registered intensity information, the terrestrial laser scanner (TLS) is able to identify efflorescence, humidity, and biodeterioration of structures, among other pathologies. It is also possible to monitor the evolution of these pathological manifestations throughout the life of the structure [8]. Some devices also have an integrated digital camera. With the help of this, it is possible to associate color information, RGB, to each point of the cloud [9]. **Figure 2** below illustrates data collection.

Thus, the point cloud is a dense representation of points in three space, forming a 3D image of the scanned structure [1, 13]. Laser scanner equipment is capable of capturing from hundreds of thousands to millions of points per second, with high accuracy and at a high acquisition rate [14]. The resolution of the point cloud, which defines the total amount and the distance between the points, i.e., the point density, can be configured by the operator before scanning. This parameter depends on the size of the smallest element to be lifted and the distance to the object [15].

Currently, some of the main manufacturers of ground laser scanners are Faro (United States), Leica (Switzerland), Riegl (Austria), Topcon (Japan), Trimble (United States), and Velodyne (United States). These companies, as well as others in the geotechnology and terrestrial survey sector, usually provide solution packages for scanning according to the buyer's design. In addition to the TLS equipment, the

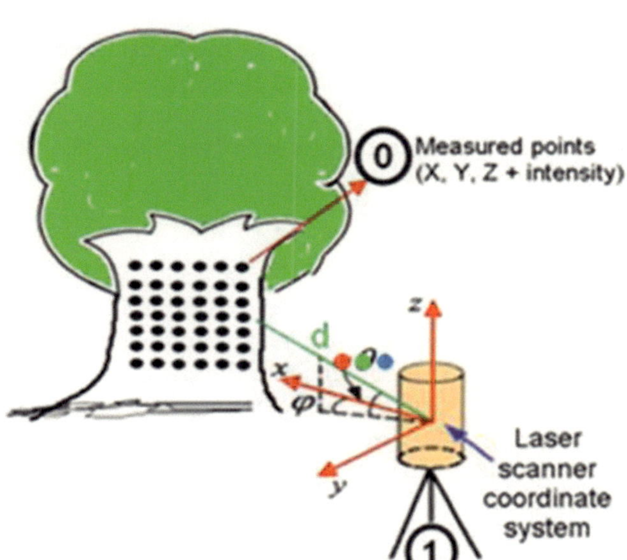

Figure 2.
Data collection with terrestrial laser scanning [9].

Manufacturer	Model	Reach	Acquisition Rate (pps)	Accuracy	Field of View (h x v)	Camera (MP)
Faro	Focus^S 350	350 m	976.000	1 mm	360° x 300°	165
Leica	RTC360	130 m	2.000.000	2.9 mm-20 m	360° x 300°	36/432
Riegl	VZ-6000	6000 m	222.000	3.5 mm-25 m	360° x 60°	5
Topcon	GLS-2000 L	500 m	120.000	4 mm-20 m	360° x 270°	5
Trimble	SX10	600 m	26.600	14 mm-100 m	360° x 300°	5
Velodyne	HDL-64E	120 m	2.200.000	2 cm	360° x 26.9°	—

Table 1.
Main features of some scanners.

necessary software for processing the data collected in the field is available, as well as technical support, as shown in **Table 1**.

In short, the main advantages of using terrestrial scanners are the speed of operation, the high precision of the generated point cloud, and the fact that the laser is not harmful to the health of the operator, especially skin and eyes. It is also possible to reach areas that are difficult to access, because it has a long range. Finally, there is the possibility of integration with GNSS systems, so that the points in the cloud can be georeferenced.

The main disadvantages are that the acquisition cost of equipment and software is still high, although the solutions currently in the market are robust and efficient. The cloud of points generated, given the high amount of information contained in it, demands high computational and graphical processing capacity, requiring the use of good processors and video cards. According to Ferraz et al. [8], external factors such as the temperature of the environment and the object to be scanned, pressure, and humidity can influence the measurements, since they affect the refraction index of the laser, changing its wavelength.

The main barriers this technology faces are the insecurity about the use of 3D environments and the insertion of this method in the traditional flow of data processing. There is still a feeling that because it is a three-dimensional cloud, the learning curve will be long. In some countries, there is little know-how of the applications and advantages of this technology.

2.3 Main applications

In recent years, several applications of Terrestrial Laser Scanning (TLS) and the point cloud generated from it have been studied by countless authors, aided in large part by the intense technological development of the last two decades. Among the application areas of TLS, they stand out in engineering and architecture, infrastructure, industry, mining and geology, archeology, ecology, and medicine and in the forensic and agricultural sectors. In engineering and architecture, the generation of as-built 2D plans, interior design, 3D modeling and integration with BIM (Building Information Modeling) technology [16], structure monitoring [17] and interventions in existing structures.

As mentioned in Refs. [2, 18], the laser scanner is used in the dimensional and structural analysis of a Spanish bridge. Pesci et al. [19] analyzed deformations and deviations of structures from TLS data. González-Jorge et al. [13] used the terrestrial laser scanning to detect the presence of biofilm and moss proliferation in reinforced concrete structures.

Teza et al. [20] developed a computational method to identify surface damage caused by loss of mass in reinforced concrete structures. Rabah et al. [21] used the TLS to detect and automatically map cracks in concrete surfaces. Brandão [22] studied its application to buildings.

In topography, it has great utility in planialtimetric topographic surveys, calculation of volumes, maps and generation of topographic profiles, and Digital Elevation Models—DEM [23].

In this case, the objective was to determine areas at risk of flooding in the urban zone. Wutke [1] mentions application in geotechnics and geophysical modeling and also mentions studies in reverse engineering, prototyping, aeronautical, vehicles, vessels, and large objects in general modeling. It is also used in the documentation of industrial plants, in mining, with calculation of ore volume, in studies of natural risks, and in verification of underbreak and overbreak during tunnel excavation [24]. Fekete et al. [25] used the TLS to scan tunnels in the process of excavation and drilling. There are also studies on the documentation and preservation of historical heritage [19, 26, 27]. In the forensic sector [28], it is used to reconstruct crime scenes.

According to Giongo et al. [5], the TLS can be used in coastal planning, flood risk assessment, telecommunications, and energy transmission networks. It is also possible to use it in agriculture [29], mining [30], mineralogy [31], and vegetation studies [32]. In medicine, Dalmolin and Santos [33] mentioned the design and manufacture of prostheses and comparative studies of volume and surface textures before and after surgeries.

Mugnai et al. [27] presented a research study designed to assess the health status of a medieval bridge built on 1500 under the Medici dynasty over the river Sieve, close to Florence, using TLS to identify anomalies and deformations. Marzouk et al. [26] proposed a framework with TLS data and BIM models in order to overcome the weaknesses of the traditional methods in Egyptian Heritage called Tosson Palace. Takhirov et al. [34] used the high-definition laser scanning technology in an extensive structural assessment of historic monuments in Uzbekistan.

Shafikani et al. [35] showed an application of TLS technology for assessing the performance of bridge infrastructures, including highway embankments, bridge decks, approach slabs, abutments, and columns supported on drilled shafts. This application also studied the ground movements. Cha et al. [36] introduced a practical feasibility study of a shape information model to monitor deformation or deflection of bridge structures using TLS. Finally, Carvalho et al. [37] presented a case study of the application of the terrestrial laser scanning in the identification of structural pathologies in a viaduct.

Armesto et al. [38] developed a methodology using TLS to estimate the deformation of arches or vaults based on the symmetry of sections obtained along the vault guideline. TLS technology applied to register as-built projects can be found in Refs. [39–41]. Armesto-González et al. [42] presented a methodology to combine the technology of the terrestrial laser scanner with the techniques of digital image processing in order to study damages on stony materials that constitute historical buildings.

In addition to the Civil Engineering field, there are several other applications of TLS technology as described below among others:

- architectural restoration of iconic buildings, such as the Glass House, built in the 1950s, as well as the MASP Structure Conservation Plan, both in the city of São Paulo;
- documentation and renovation of the Malé Hukuru Miskiy, located at Malé city in the Republic of Maldives;
- based on the fire that almost destroyed Notre Dame Cathedral in Paris a lot of other Museums, Heritages, Opera Houses, and other iconic buildings are investing in this scope of work.
- volumetric measurement of chemicals products in storage sheds;
- elaboration of three-dimensional electronic model and architectural plans on a large-scale map;
- planialtimetric mine survey for mining planning and track of tunnels; and
- new experiences and services for the real estate market: accurate floorplans for customers to know exactly what they are paying for.

3. Reinforced concrete structures

3.1 Composition and characteristics of reinforced concrete

A mixture of cement and water is called a paste, while mortar is a mixture of paste with a fine aggregates. Simple concrete is defined as mortar to which a large number of aggregates have been added, i.e., fine and coarse aggregates [43]. When mixed, the constituent elements of the concrete, still fresh, form an alloy of plastic consistency that can be molded into various forms, serving various purposes. This explains its wide dissemination and use as construction material. The forms that ensure its final shape after hardening can be of wood or steel, with the latter being much more common nowadays [44].

The initial plasticity of the concrete is of fundamental importance for the handling, transport, and pouring in the forms. After curing, it acquires consistency and cohesion, due to the chemical reactions between the components mentioned above [43].

Simple concrete is an element formed by agglomerate (cement, composed of a mixture of limestone and clay, called clinker), water, small aggregate (sand), and large aggregate (gravel or stone). Additionally, additives can be used, among them accelerators and hardening retarders, plasticizers and superplasticizers, viscosity modifiers, air incorporators, and colorants, which aim to change the concrete's characteristics according to the use and purpose foreseen in the project [45]. Cement, the main material used in the mixture, can also contain additions to it is basic composition, such as fly ash, pozzolan, and active silica. Poor quality of the materials that make up the concrete can negatively influence its properties [2].

The main advantages of the use of concrete are that the constituent materials are economically accessible and abundant, especially in Brazil, and the facility of molding and assembly. In addition, it has high resistance to fire, stability under the action of weather, and resistance to shocks and vibrations. However, some of the main disadvantages are the high weight (approximately 2.5 metric tons per cubic meter), the low thermal and sound protection, the development of cracks, and the requirement for constructive rigor and support apparatus, such as shoring, thickening, and curing processes [46].

The association of simple concrete with steel bars, forming reinforced concrete, occurs because while concrete presents high compression resistance, steel has high tensile resistance [2]. According to Gonçalves [47], reinforced concrete is composed of simple concrete, steel reinforcement, and a component to assure good adhesion between these two materials.

The use of concrete with steel is only possible due to the joint work between these two materials, given by the adherence between them. Additionally, its dilation coefficients are similar, so the behavior of concrete under temperature variations is compatible with that of steel. In addition, concrete acts as a double barrier for steel, physically and chemically. On the one hand, the concrete separates and protects the steel from the external environment through the covering layer. On the other hand, the concrete protects the steel against oxidation, through an alkaline environment formed during the handling period of the concrete. This chemically inhibiting layer is called a passive film. The aqueous solution consists mainly of hydroxyl ions (OH^-), which causes the high alkalinity of concrete [48, 49].

The Brazilian Standard (NBR) 6118, the main Brazilian reference for the design of concrete structures, defines reinforced concrete structures as those whose structural behavior is conditioned by the adherence between concrete and reinforcement. In this type of structure, initial elongations of reinforcement are not applied before the materialization of this adherence [50–52].

3.2 Pathologies in concrete structures

The fast development of civil construction to meet a growing demand for buildings of various types has promoted a great scientific and technological leap. But many structures show unsatisfactory performance due to aspects such as involuntary failures, imperfections, misuse of materials, natural aging and errors in designs, and execution [53].

The study of pathologies of concrete structures and construction in general has been growing in recent years. Many people believed that concrete structures would be eternal, requiring no maintenance or intervention, even when built without specific care and exposed to aggressive environments. However, several cases of degradation and collapse, sometimes early, occurred from the 1980s on, which exposed the risks to safety and caused high maintenance costs, promoted a change in this perspective, with an understanding of the importance of preventing degradation and performing preventive and corrective maintenance [54–56].

Materials for construction are subject to natural wear and aging processes, usually accelerated by environmental factors. Several agents are responsible for the uninterrupted degradation of concrete structures [57]. Ideally, this happens at a predictable rate, so that the structure maintains its functional capacity during its useful life. In cases where deterioration occurs at higher rates than predicted, intervention in the structure may be necessary [58].

Thus, the action of the environment together with failures of design, execution, and/or maintenance leads to the appearance of pathological manifestations in structures. The pathologies are the main problems that compromise the useful life of concrete structures [53].

In this work, with the objective of better presenting and conceptualizing the various types of structural pathologies, the classification defined by Pase [45] is used. According to this author, mechanical processes are related to pathologies caused by accidents, overloads, or impacts. Physical processes are freeze-thaw cycles, thermal stresses, deformations due to shrinkage, or creep and surface wear due to abrasion, erosion, and cavitation. Chemical processes are attacked by acids, sulfates, chlorides, alkali-aggregate reaction, and leaching. Electrochemical processes are carbonation and corrosion of steel. Finally, biological processes are those caused by vegetation growing on structures and within their cracks.

The process of treating pathological manifestations in concrete structures can be divided into three distinct phases. Initially, the pathologies that act on the structure are identified through visual inspections and tests on-site or in the laboratory (through the extraction of specimens). Then, a diagnosis of these pathologies is made, with the objective of determining their causes and mechanisms of action. Finally, from the information of the first two steps, an intervention program is defined, whose function is to specify the actions that must be taken to mitigate such pathologies. These actions include the choice of methods and materials to be used in the recovery of the structure. In some cases, depending on the type and degree of progress of the deterioration, it may also be necessary to perform structural reinforcement in order to reestablish the cross-sectional and resistance capacity.

4. Methodology

This section presents and comments on the results of the evaluations of the normalized point cloud of the São Cristóvão Viaduct, as well as the three-dimensional model prepared from it and the classification made. Thus, the objective was to determine the possibility of identifying these pathologies through the point cloud and these products, shown below.

4.1 General characteristics of the structure studied

The structure chosen for analysis was the “Old” São Cristóvão Viaduct, located in the district of the same name in the northern zone of the city of Rio de Janeiro. The viaduct overpasses train and metro lines and is located next to the Oduvaldo Cozzi Viaduct.

The viaduct is made of reinforced concrete. Its structural arrangement comprises a total of six longitudinal beams along its length. These beams receive the forces coming from the slab over which vehicles circulate. The structure is segmented into six spans, one on the side of the Maracanã district (access), one above the Rio Metro railway line (aboveground extension of the subway system), two above the Supervia railway line (metropolitan commuter trains), one above General Herculano Gomes Street, and one on the way down the viaduct, next to Quinta da Boa Vista Park. Each span is separated from the adjacent one by a transversal line of six pillars each, which bear the forces from the six longitudinal beams. Additionally, each span has five transversal beams (perpendicular to the longitudinal beams), which also receives loads from the slabs, as shown in **Figure 3**.

DOI: http://dx.doi.org/10.5772/intechopen.1001043

Figure 3.
Structural arrangement of São Cristóvão viaduct.

4.2 Field survey

The survey took place on May 27, 2019, and lasted approximately 2 hours. According to the geometry of the viaduct structure, the positioning of the station platforms in relation to it, the fact that the station is extremely busy and the staff team of the concessionaire did not allow us to set up the equipment over the train lines for mapping, 14 equipment position points (setup points) were determined. Additionally, a fifteenth point was defined outside the station, on the sidewalk of General Herculano Gomes Street, in order to generate a point cloud in this region of the viaduct. Thus, the survey generated 15 scenes, as shown in **Figure 4**, that is, 15 different point clouds, one for each point where the device was positioned.

In this work, we chose to use the device with a medium acquisition rate. In this way, the taking of photographs in each scene lasted approximately 3 minutes. The scanning of each scene lasted approximately 4 minutes, totaling approximately 7 minutes for each position point. This configuration took place in this matter in order to acquire data from the segments between adjacent longitudinal beams, given that they were too close to each other.

4.3 Data processing

At the end of the first scan, the device was taken to the second position point. The scanning process was then repeated for all 15 points where the device was positioned.

Finally, with all the clouds generated, the raw data were processed in the university office. This can be done by any computer that has the minimum computational requirements defined by the equipment manufacturer. In this work, we used a computer with a 6-core processor, 12 MB cache, and maximum frequency of 3.8 GHz, along with a graphics card (GPU) of 1GB and 32 GB RAM.

The Leica Cyclone Register 360 software was used for the refinement and completion of the scenes. When the program was opened, a new project was created, and then, the 15 cloud files were imported. These files are native to the BLK 360 terrestrial laser scanner and have *.blk extension. This way, it was only possible to open them

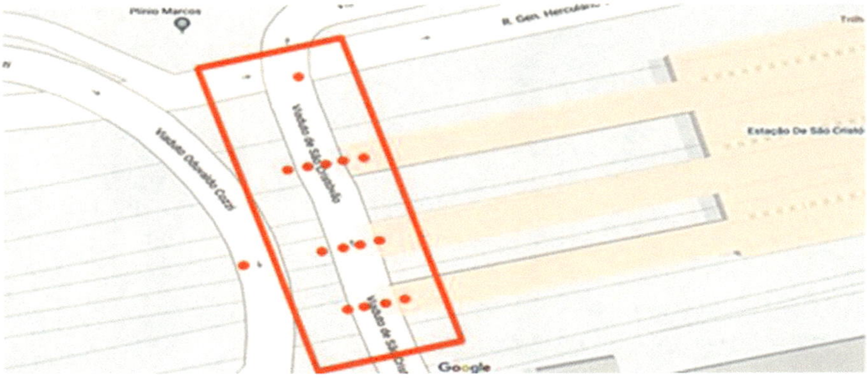

Figure 4.
Schematic representation of the 15 position points of the equipment (Google maps, 2019).

with the Leica proprietary software, from which they are exported to other formats. In this work, each cloud file had approximately 354 MB, totaling approximately 5.31 GB for the 15 files.

In this work, the results of the Register were an error of 0.006 meters, overlap of 44% between "clouds", force of 67%. In this case, "cloud" by "cloud" was employed as the registration method.

The scenes were then recorded between pairs of point clouds, as shown in **Figure 3(a)**, where each green line indicates a connected and recorded pair. Each of these connections was called a link. For the 15 clouds in this project, 26 links were established.

After creating the 26 links, the process of recording point clouds was completed. From this, the software displayed the recording quality parameters: accuracy (bundle error), overlap, and strength, according to **Figure 3(b)**, and issued a report of the process. At the end of the recording, the consolidated cloud was exported to the open format *.pts. The file was approximately 5.76 GB.

Thus, after all the scenes were recorded, the cloud was reduced so that it only contained points relative to the São Cristóvão Viaduct. This procedure was performed in the CloudCompare program (version 2.10.3 Zephyrus). This is free software for processing three-dimensional point clouds under the GNU General Public License. This reduction consisted of selecting the regions that did not correspond to the studied structure and eliminating them from the cloud, as can be seen in **Figure 5(c)**, in red.

4.4 Consolidated point cloud

After recording the clouds, a single consolidated cloud was obtained, which presented all the elements in the environment around the viaduct, captured by the terrestrial laser scanner during field work.

The consolidated cloud, as can be seen in **Figure 6**, although visually attractive, detailed, and rich in elements, presented a number of features far superior to the object of study. In this figure, it is possible to see that, besides the structure of the viaduct, data from the station platforms, the nearby Oduvaldo Cozzi Viaduct, as well as the nearby vegetation, passengers and the Supervia train cars themselves were also captured.

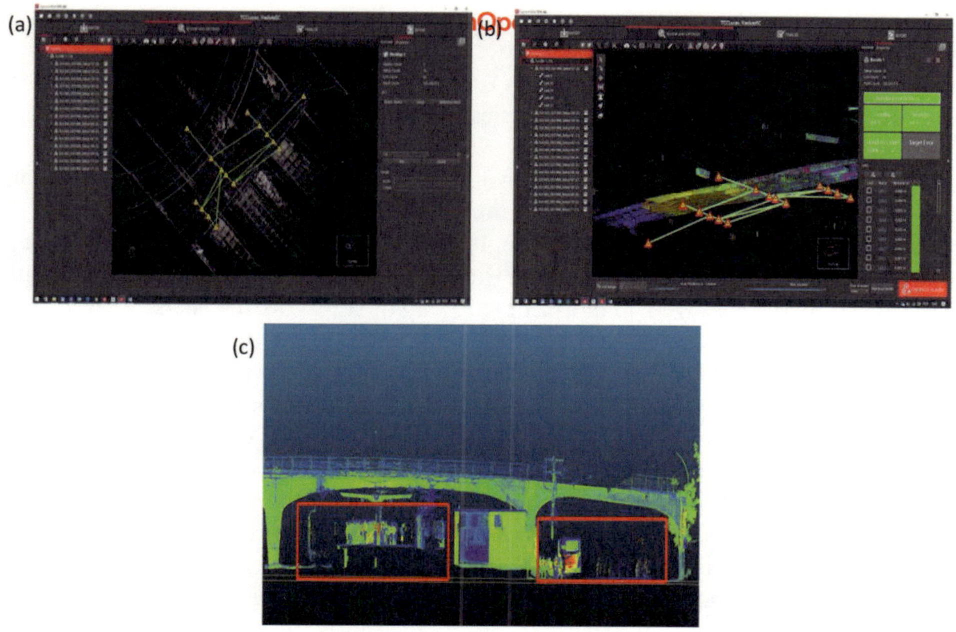

Figure 5.
Legend: (a) scene registration, (b) conclusion of the cloud recording process, (c) elimination of features in CloudCompare software.

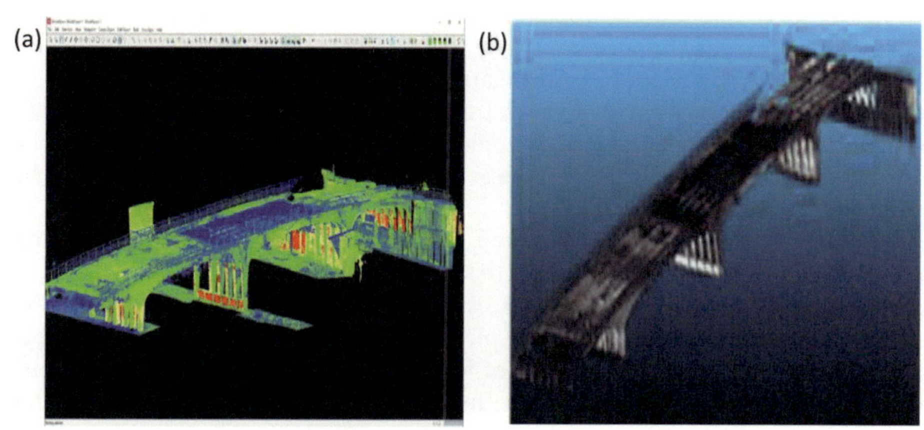

Figure 6.
Legend: (a) point cloud after recording and cleaning. (b) Normalized point cloud, in grayscale.

5. Results and discussion

This section presents in greater detail the normalized point cloud and the three-dimensional surface generated for the São Cristóvão Viaduct, in order to identify structural pathologies. For this purpose, the viaduct was segmented into four sections defined according to the position of the pillar lines. For each section, the point cloud, the three-dimensional surface and photos of the corresponding region are shown, in order to establish a parallel between the visually identified pathologies and their representations in the products generated from the terrestrial laser scanner survey.

This comparison aims to study the applicability of this technique for the identification and cataloging of pathological manifestations.

The pathologies identified in sections 1, 3, and 4 during this study are available from the corresponding author upon request. We present below the pathologies identified in Section 2.

Both the point cloud and three-dimensional surface present continuous behavior, mainly in relation to Section 1. This continuity can be explained by the location of more points for the survey of this section. It is possible to notice a lower occurrence of shadow regions, which stand out near the pillars. Shadows, although less frequent, were caused by the passage and constant stopping of trains during scans, also acting as a physical barrier to the integral capture of the structure of this stretch.

Failures can also be noticed in the external regions of the viaduct, notably at the edges and in the upper portion, near the slab. This behavior is explained by the positioning of the scanner during the scans. As the survey was done from the station platforms, and since the coverage of these platforms extended to the vicinity of the viaduct along the platforms, the field of view to the top of the viaduct was restricted. Also contributing to this failure is the fact that the coverage of the upper part of the viaduct slab was not part of the survey.

The pillars show good representation in the point cloud, given the fact they are located in a region closer to the scanner, thus being hit by the beams orthogonally. The bottom of the slab as well as the sides and bottom of the beams also have good representation in the cloud.

The three-dimensional surface is consistent with the point cloud, showing discontinuities in regions with less point density. Slight presence of noise also can be observed, more frequent in the upper part of the slab, where the open surface and the scarcity of points made the creation of triangular surfaces difficult.

The texture of the surface is uniform and consistent with the structure it represents, with good coverage of the slab and beams. The pillars have accurate representation, consistent with their geometry, including the application of colors to their surface. However, atypical roughness is observed in the external pillars of this section.

Some critical regions can be observed in this section, as seen in the details of the point cloud, **Figure 7(a)**, **(c)**, and **(e)**, and the three-dimensional surface, **Figure 5(b)**, **(d)**, and **(f)**. There are moisture stains along the edges of the slab, at the end of the viaduct, which is also explained by the fact that it is a region more subject to infiltration and percolation of water. These can be better observed in the normalized point cloud than in the three-dimensional surface, which can be explained by the difficulty in creating them. Moisture is also observed in some internal regions of the viaduct slab.

Analysis of the point cloud also shows the existence of slabs and beams in the regions of the structure with exposure of shallow reinforcement (rebar's) on the concrete surface. With the gray scale coloration of the cloud, these regions can be more easily discerned.

Through analysis of the three-dimensional surface, although it is possible to verify the presence of moisture stains and rebar exposure, its identification cannot be done easily and in isolation. However, it is possible to identify the regions where biodeterioration has occurred.

In this section, the slab and beams of the viaduct are more affected, with greater presence of pathological manifestations than the pillars, which in general do not show structural deterioration.

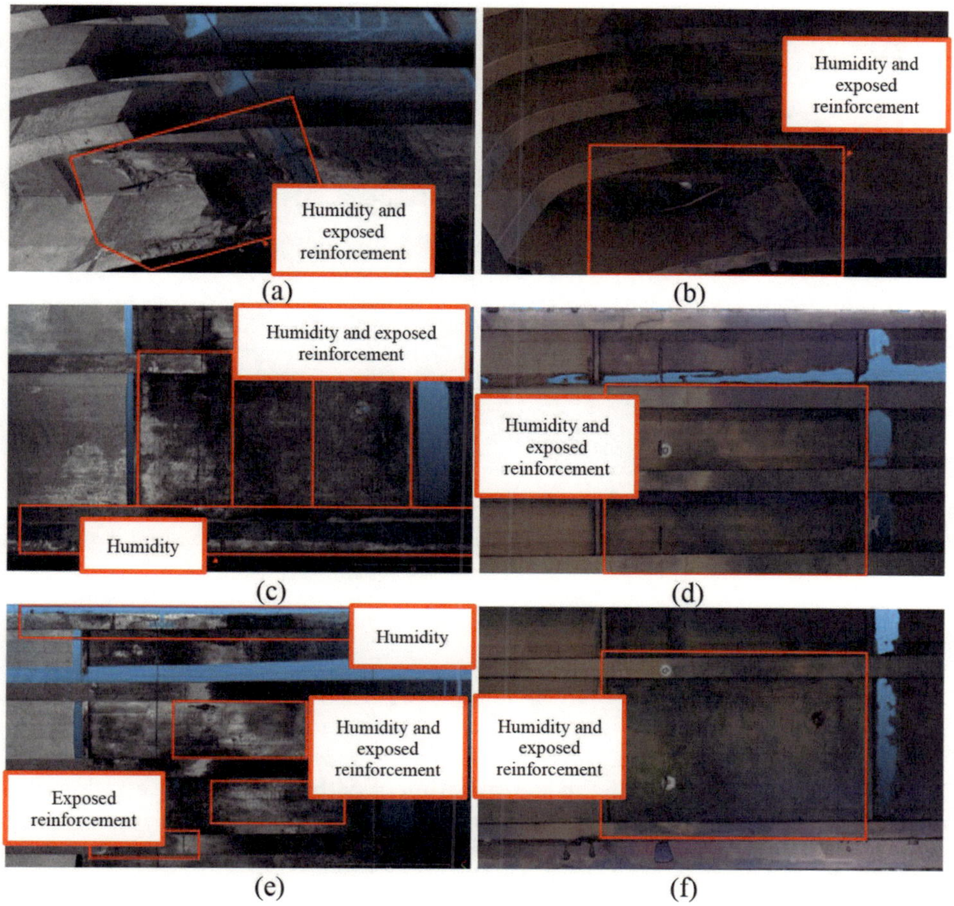

Figure 7.
Pathologies presents in the viaduct. (a), (c), and (e) Gray scalar; (b), (d), and (f) real color.

Additionally, besides the analysis of the standardized point cloud and the three-dimensional surface generated from it, we also did a point cloud classification based on the intensity of the laser return combined with the shape of the structure surface. These classes were defined by using the plugin/algorithm "qCANUPO" from the CloudCompare software and selecting samples from the cloud itself and then applying them to the points in their entirety.

This classification aimed to propose a semi-automatic method for identifying structural pathologies from the point cloud. This method allows quicker field work, and it is especially helpful in situations where people cannot get access to the structure entirely. Since this is an initial study, for the cloud generated in this work, two classes were defined so it would be possible to differentiate parts of a viaduct, one for healthy regions, identified by the color red, and another for deteriorated regions, identified by blue, according to **Figure 8**.

In future studies, it is recommended that the unhealthy class be split into more sub-categories, according to the pathologies found in the structure. This will allow a more assertive global analysis, enabling actions more targeted to each pathology to be taken. In addition, this more segmented analysis can also allow the correct quantification of the areas affected by each pathology.

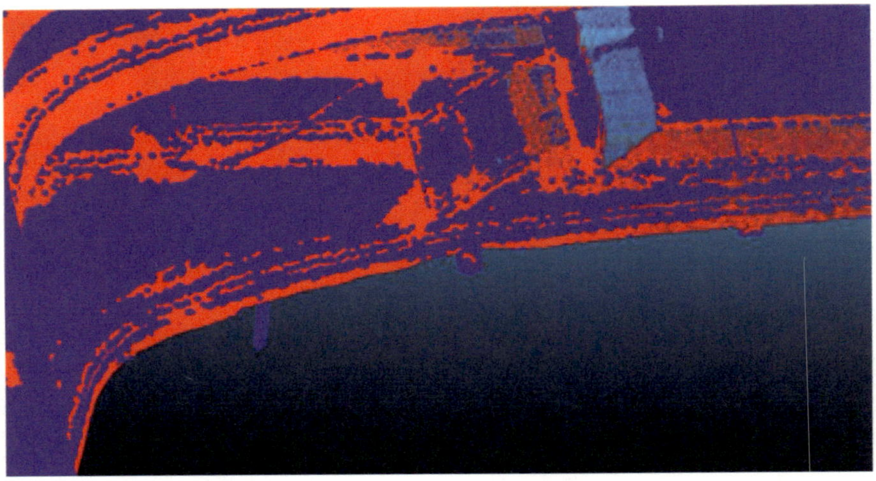

Figure 8.
Classification of the cloud for part of section 2. Legend: healthy regions (color red) and deteriorated regions (color blue).

6. Conclusion

Nowadays, bridge and viaduct inspections are performed visually using a camera. Thus, the presence of damage is not an accurate quantitative evaluation, depending exclusively on the inspector's qualitative criteria, as well as productivity is low. The use of the TLS technique in the identification of pathological manifestations in bridges and viaducts can be incorporated into that inspection activity to complement the existing methodology.

The TLS is capable of detecting biodeterioration, humidity, cracks, and efflorescence, as well as allowing the viewing of dimensional changes and deformations in the order of centimeters. By enabling the visualization and scanning of a structure without direct contact with it, TLS would be an option for inspections of bridges and viaducts whose pillars, beams, and board were inaccessible to the inspector's contact.

Terrestrial laser scanning (TLS) has application in several areas of knowledge, notably in engineering. Given the characteristics of the data that this equipment generates, forming a point cloud, it is possible to formulate several spatial products and perform qualitative and quantitative analysis. One application still relatively little explored is its use to identify pathologies in reinforced concrete structures. This is done by scanning the structure of interest and using the intensity and color data of the return pulse collected by the scanner to identify healthy and deteriorated regions. Thus, this work aimed to study the applicability of this equipment in civil engineering, to support maintenance programs of buildings and other structures.

Based on the bibliographic research of the theoretical bases that underpin TLS, its operating principles, characteristics, and application present in the literature, as well as concrete structures, their types, composition, and main degradation processes, it was possible to confirm the theoretical relation between the two themes. In order to deepen the study of this relation, we presented a case study of a viaduct made of reinforced concrete.

By analyzing each point cloud after the elimination of features, it was possible to observe good general representation of the viaduct structure and critical areas to be used for structural analysis. In some regions, however, especially in sections 1 and 4, significant discontinuities in the cloud and shadow regions, caused by the presence of physical barriers at the time of the survey, were noted. These barriers consisted of partition walls and the presence of moving trains, people, and vehicles during the scanning. Although this did not make the work impossible, such barriers and impediments resulted in areas with sparse data, so the overall analysis of the structure was impaired.

The use of the terrestrial laser scanning for the identification of structural pathologies is a clean alternative for monitoring and inspection of concrete structures. Compared to other existing approaches, it is a nondestructive and noninvasive method that where, besides not generating damage to the evaluated structure, does not produce any waste to be discarded in the environment and can guarantee a trustworthy result of the collected and processed data. Moreover, the most relevant characteristic of the technology is the great speed and quality in data collection. In general, this technique enables the evaluation of pathologies in the structure to be improved quantitatively and qualitatively. Thus, better evaluations result in more corrective maintenance and major repairs, with the benefit of an extended lifetime and greater safety.

Acknowledgements

All the authors are acknowledging to Rio de Janeiro Municipal Government, through its General Project Coordination Office (CGP), part of Municipal Department of Infrastructure and Housing (SMIH), to authorize the use of the viaduct for an object of analysis. Also, the authors acknowledge the Rio Metro railway line and Supervia railway line, to use the platform to install the terrestrial laser scanning.

Compliance with ethical standards

The authors declare that they have no conflict of interest.

Authors with contributor roles

Sergio Orlando Antoun Netto, DSc. (Conceptualization: Equal; Methodology: Supporting; Resources: Equal; Supervision: Lead; Writing—review & editing: Lead).

Lucas Pires Chagas Ferreira de Carvalho, Engineer (Conceptualization: Lead; Formal Analysis: Lead; Investigation: Lead; Methodology: Supporting; Writing—original draft: Equal).

Ana Waldila de Queiroz Ramiro Reis, M.Sc. (Formal Analysis: Lead; Investigation: Lead; Methodology: Supporting; Writing—original draft: Lead).

Leonardo Vieira Barbalho, M.Sc. (Data curation: Lead; Methodology: Supporting; Software: Supporting; Validation: Supporting).

Lucas de Campos Rodrigues, Engineer (Methodology: Supporting; Resources: Lead; Software: Lead; Visualization: Supporting).

Additional information

This chapter has previously been published as a preprint: Netto SOA, de Carvalho LPCF, Reis AW de QR, Barbalho LV, Rodrigues L de C. Study of the Application of Terrestrial Laser Scanning for Identification of Pathologies in Concrete Structures [Internet]. Research Square Platform LLC; 2021. Available from: http://dx.doi.org/10.21203/rs.3.rs-1001215/v1

Author details

Sérgio Orlando Antoun Netto[1*], Lucas Pires Chagas Ferreira de Carvalho[1], Ana Waldila de Queiroz Ramiro Reis[2], Leonardo Vieira Barbalho[1] and Lucas de Campos Rodrigues[1]

1 Engineering Faculty, Department of Cartography, State University of Rio de Janeiro, Rio de Janeiro, Brazil

2 Engineering Faculty, Department of Civil, State University of Rio de Janeiro, Rio de Janeiro, Brazil

*Address all correspondence to: sergioantoun@gmail.com

References

[1] Wutke JD. Métodos para avaliação de um Sistema laser scanner terrestre. Curitiba: 86 f. Dissertação (Mestrado) - Curso de Pós-Graduação em Ciências Geodésicas, Departamento de Geomática, Universidade Federal do Paraná; 2006

[2] Pavi S, Bordin F, Veronéz MR. O potencial do uso do laser scanner terrestre para a identificação de manifestações patológicas em obras de arte especiais: uma revisão bibliográfica. X Congreso Internacional sobre Patologia y Recuperación de Estruturas - CINPAR, Santiago de Chile, Anais...Santiago de Chile; 2014. pp. 1-15

[3] Fitz PR. Cartografia básica. 2nd ed. São Paulo: Oficina de Textos; 2008. p. 144

[4] Florenzano TG. Imagens de satélite para estudos ambientais. 1st ed. São Paulo: Oficina de Textos; 2002. p. 97

[5] Giongo M et al. Lidar: princípios e aplicações florestais. Pesquisa Florestal Brasileira. 2010;**30**(63):231-244

[6] Longley PA et al. Sistemas e ciência da informação geográfica. 3rd ed. Porto Alegre: Bookman; 2013. p. 529

[7] Nascimento Júnior JO do. Schuler CA, Cavalcante RB, Anais de S. III Simpósio Regional de Geoprocessamento e Sensoriamento Remoto Aracaju/SE, 25 a 27 de outubro de. 2006

[8] Ferraz R da S, Souza SF de, Reis MLL. Laser Scanner Terrestre: teoria, aplicações e prática. Revista Brasileira de Geomática, Pato Branco. 2016;**4**(2):99-109

[9] Bordin F. Laser scanner terrestre na caracterização de alvos florestais. In: Tese (Doutorado) – Programa de Pós-Graduação em Sensoriamento Remoto. Porto Alegre: Universidade Federal do Rio Grande do Sul; 2015. p. 89 f

[10] Brandalize AA. Perfilamento a LASER: comparação com métodos fotogramétricos. Curitiba: Esteio Engenharia e Aerolevantamentos S.A; 2001

[11] Wehr A, Lohr U. Airborne laser scanning – An introduction and overview. ISPRS Journal of Photogrammetric and Remote Sensing. 1999;**54**:68-82

[12] Lenartovicz IR. Avaliação da potencialidade do laser scanner terrestre no monitoramento de estruturas: estudo de caso UHE Mauá. 121 f. Dissertação (Mestrado) - Programa de Pós-Graduação em Ciências Geodésicas. Curitiba: Departamento de Geomática, Universidade Federal do Paraná; 2013

[13] González-Jorge H et al. Monitoring biological crusts in civil engineering structures using intensity data from terrestrial laser scanners. Construction and Building Materials. 2012;**31**:119-128

[14] Bellian JA, Kerans C, Jennette DC. Digital outcrop models: Applications of terrestrial scanning lidar technology in stratigraphic modeling. Journal of Sedimentary Research. 2005;**75**(2):166-176

[15] Tommaselli AMG. Um Estudo sobre as Técnicas de Varredura a Laser e Fotogrametria para Levantamentos 3D a Curta Distância. São Paulo, UNESP Presidente Prudente: Geodésia Online; 2003

[16] Pu S, Vosselman G. Knowledge based reconstruction of building models from

terrestrial laser scanning data. ISPRS Journal Photogrammetry and Remote Sensing. 2009;**64**(6):575-584

[17] Godycka KN, Szulwic J, Ziółkowski P. The Method of Analysis of Damage in Reinforced Concrete Beams Using Terrestrial Laser Scanning. Poland: Faculty of Civil and Environmental Engineering, Gdansk University of Technology; 2014 14th International Multidisciplinary Scientific GeoConference SGEM

[18] Lubowiecka I. Historic bridge modelling using laser scanning, ground penetrating radar and finite element methods in the context of structural dynamics. Engineering Structures. 2009;**31**(11):2667-2676

[19] Pesci A, Casula G, Boschi E. Laser scanning the Garisenda and Asinelli towers in Bologna (Italy): Detailed deformation patterns of two ancient leaning buildings. Journal of Cultural Heritage. 2011;**12**(2):117-127

[20] Teza G, Galgaro A, Moro F. Contactless recognition of concrete surface damage from laser scanning and curvature computation. NDT&E International. 2009;**42**(4):240-249

[21] Rabah M, Elhattab A, Fayad A. Automatic concrete cracks detection and mapping of terrestrial laser scan data. NRIAG Journal of Astronomy and Geophysics. 2013;**2**:6. Available from: http://www.riegl.com/nc/products/terrestrialscanning/produktdetail/product/scanner/33 [Accessed: September 14, 2019]

[22] Brandão AM da S. Qualidade e durabilidade das estruturas de concreto armado. 137 f. Dissertação (Mestrado) - Curso de Engenharia de Estruturas. São Carlos: Universidade de São Paulo; 1998

[23] Meouche RE et al. Using a laser scanning to construct a 3D numerical model to study the flooding risk in urban area. IACSIT International Journal of Engineering and Technology. 2013;**5**(3):7

[24] Gonçales R. Dispositivo de varredura laser 3D terrestre e suas aplicações na engenharia, com ênfase em túneis,103 f. Dissertação (Mestrado) - Programa de Pós-graduação em Engenharia de Transportes. São Paulo: Departamento de Engenharia de Transportes, Universidade de São Paulo; 2007

[25] Fekete S, Diederichs M, Lato M. Geotechnical and operational application for 3-dimensional laser scanning in drill and blast tunnels. Tunnelling and Underground Space Technology. 2010;**25**(5):614-628

[26] Marzouk M, Metawie M, ElSharkawy M, Eid A, Hawas S. Application of laser scanning technology in energy analysis and structural health monitoring of heritage buildings. In: CSCE Annual Conference Growing with Youth – Croître avec les jeunes At: Laval (Greater Montreal) Project: Framework for Restoration of Heritage Using Building Information Modeling (FresHBIM) June 12-15, 2019

[27] Mugnai F, Lombardi L, Tucci G, Nocentini M, Gigli G, Fanti R. Geomatics in bridge structural health monitoring, integrating terrestrial laser scanning techniques and geotechnical inspections on a high value cultural heritage. International Archives of the Photogrammetry, Remote Sensing & Spatial Information Sciences. 2019;**XLII-2/W11**:895-900

[28] Sansoni G, Cattaneo C, Trebeschi M, Gibelli D, Poppa P, Porta D, et al. Scene of-crime analysis by a 3-dimensional optical digitizer: A useful perspective for forensic science. The American journal

of forensic medicine and pathology. 2011;**5**(2):280-286

[29] Lumme J, Karjalainen M, Kaartinen H, Kukko A, Hyyppä J, Hyyppä H, et al. Terrestrial laser scanning of agricultural crops. The International Archives of the Photogrammetry, Remote Sensing and Spatial Information Sciences. 2008;**XXXVII**(Part B5):563-566

[30] Barchik EAL, Moser I, Santos DS, Martins BD. Aplicação do scanner terrestre ILRIS-3D no ramo da mineração. In: Simpósio Brasileiro de Sensoriamento Remoto, 13. (SBSR). Florianópolis. Anais... São José dos Campos: INPE; 2007. pp. 3631-3635

[31] Kurz TH, Buckley SJ, Howell JA. Close range hyperspectral imaging integrated with terrestrial lidar scanning applied to rock characterization at centimetre scale. International Archives of the Photogrammetry, Remote Sensing and Spatial Information Sciences. 2012;**XXXIX-B5**:417-422. DOI: 10.5194/isprsarchives-XXXIX-B5-417-2012

[32] Lingnau C, Silva M, Santos D, Machado LJ. Mensuração de alturas de árvores individuais a partir de dados laser terrestre. Revista Ambiência, Guarapuava. 2008;**4**(1):85-96

[33] Dalmolin Q, Santos DR dos. Sistema Laser Scanner: Conceitos e princípios de funcionamento. 3rd ed. Curitiba: UFPR; 2004. p. 97

[34] Takhirov S, Gilani A, Quigley B, Myagkova L. Structural health monitoring and assessment of seismic vulnerability of historic monuments on the Great Silk Road based on laser scanning. Em: Aguilar R, Torrealva D, Moreira S, Pando MA, Ramos LF, editors. Structural Analysis of Historical Constructions. RILEM Bookseries, Vol. 18. Cham: Springer; 2019

[35] Shafikani A, Bheemasetti T, Puppala A. Performance evaluation of a bridge infrastructure using terrestrial laser scanning technology. Geotechnical Testing Journal. 2019;**42**(6):1587-1605. DOI: 10.1520/GTJ20170325

[36] Cha G, Park S, Oh T. A terrestrial LiDAR-based detection of shape deformation for maintenance of bridge structures. Journal of Construction Engineering and Management. 2019;**145**:04019075. DOI: 10.1061/(ASCE)CO.1943-7862.0001701

[37] Carvalho LPCF de. Estudo da aplicação do laser scanner terrestre na identificação de patologias em estruturas de concreto. 153 f. Projeto final (Graduação em Engenharia Cartográfica) – Faculdade de Engenharia. Rio de Janeiro: Universidade do Estado do Rio de Janeiro; 2019

[38] Armesto J, Roca-Pardiñas J, Lorenzo H, Arias P. Modelling masonry arches shape using terrestrial laser scanning data and nonparametric methods. Engineering Structures. 2010;**32**(2):607-615

[39] Tang P, Huber D, Akinci B, Lipman R, Lytle A. Automatic reconstruction of as-built building information models from laser-scanned point clouds: A review of related techniques. Automation in Construction. 2010;**19**(7):829-843

[40] Brilakis I, Lourakis M, Sacks R, Savarese S, Christodoulou S, Teizer J, et al. Toward automated generation of parametric BIMs based on hybrid video and laser scanning data. Advanced Engineering Informatics. 2010;**24**(4):456-465

[41] Bosché F. Automated recognition of 3D CAD model objects in laser scans and calculation of as-built dimensions

for dimensional compliance control in construction. Advanced Engineering Informatics. 2010;**24**(1):107-118

[42] Armesto-González J, Riveiro-Rodrigues B, González-Aguilera D, Rivas-Brea T. Terrestrial laser scanning intensity data applied to damage detection for historical buildings. Journal of Archaeologic Armesto-González al Science. 2010;**37**(12):3037-3047

[43] Almeida LC de. Notas de aula da disciplina Estruturas IV – Concreto armado. Departamento de Estruturas, Faculdade de Engenharia Civil (FEC) da Universidade Estadual de Campinas (UNICAMP). 2002

[44] Andolfato RP. Controle tecnológico básico do concreto. Núcleo de Ensino e Pesquisa da Alvenaria Estrutural (NEPAE) da Universidade Estadual Paulista (UNESP), Ilha Solteira. 2002

[45] Pase MC. Manifestações patológicas em estruturas de concreto armado (2017). 57 f. TCC (Graduação) - Curso de Engenharia Civil. Santa Maria: Universidade Federal de Santa Maria; 2017

[46] Couto JAS et al. O concreto como material de construção. Caderno de Graduação - Ciências Exatas e Tecnológicas. Sergipe. 2013;**1**(17):49-58

[47] Gonçalves EAB. Estudo de Patologias e suas Causas nas Estruturas de Concreto Armado de Obras de Edificações. 174 f. In: Monografia (Graduação em Engenharia Civil). UFRJ – Universidade Federal do Rio de Janeiro. Escola Politécnica. Rio de Janeiro: UFRJ; 2015

[48] Rincón IT et al. Manual de inspeccion, evaluacion y diagnostico de corrosion en estructuras de hormigon armado. 1997

[49] Souza Júnior TF de. Tecnologia e qualidade do material concreto nas construções agroindustriais. 2004. 215 f. Dissertação (Mestrado) - Curso de Engenharia Agrícola. Minas Gerais: Universidade Federal de Lavras. p. 2004

[50] Associação Brasileira de Normas Técnicas. Disponível em: <http://www.abnt.org.br/>. [Acesso em: out. 10, 2019]

[51] Associação Brasileira de Normas Técnicas (ABNT). NBR 6118: Projeto de estruturas de concreto - Procedimento. Rio de Janeiro; 2014

[52] Associação Brasileira de Normas Técnicas (ABNT). NBR 7480: Aço destinado a armaduras para estruturas de concreto armado - Especificações. Rio de Janeiro; 2007

[53] Ambrósio T da SP. Tratamento e Reforço de Estruturas de Concreto no Metrô de São Paulo. 2004. 130 f. São Paulo: TCC (Graduação) - Curso de Engenharia Civil, Universidade Anhembi Morumbi; 2004

[54] Bastos HC do N, Miranda MZ. Principais patologias em estruturas de concreto de pontes e viadutos: manuseio e manutenção de obras em arte especial. Revista Construindo, Belo Horizonte. 2017;**9**:93-101

[55] Bertolini L. Materiais de construção – patologia, reabilitação, prevenção. 1st ed. São Paulo: Oficina de Textos; 2010. p. 414

[56] Cavalcante FSS. Generalidades sobre a corrosão em estruturas de concreto armado. 2009. 57 f. Trabalho de Conclusão de Curso (Graduação) – Curso de Engenharia Civil. Sobral: Universidade Estadual Vale do Acaraú; 2009

[57] Santos BO de C, Interlandi C. Estruturas de Concreto Armado: Patologias e suas Consequências Estudo

de Caso. Engenharia Estudo e Pesquisa, São Paulo. 2016;**16**(2):4047

[58] Castro EK de. Desenvolvimento de Metodologia para Manutenção de Estruturas de Concreto Armado. 1994. 129 f. Dissertação (Mestrado) - Curso de Engenharia Civil. Brasília: Universidade de Brasília; 1994

Chapter 13

The Effect of Transverse Reinforcement Corrosion on the Axial Bearing Capacity of Reinforced Concrete Columns

Ali Goharrokhi

Abstract

Confined transverse reinforcements play important roles in improving the strength and ductility of reinforced concrete columns so that this fact is mentioned in most design codes. Chloride corrosion of such reinforcements has seriously become a growing challenge for the durability of structures leading to reductions in the capacity of reinforced concrete structures and their lifetime. Because the transverse reinforcements have small concrete covers and close to free surfaces, the corrosion in these rebars occurs earlier and more severe than longitudinal reinforcements. Reductions in confinement, deformation capacity, and ductility of reinforced concrete structures are the major consequences of corrosion. Therefore, the main objective of this article is to investigate different reinforced concrete columns under diverse corrosion levels.

Keywords: reinforced concrete columns, transverse reinforcement, corrosion, confinement, axial bearing capacity

1. Introduction

Columns are one of the main critical elements of a civil structure that play vital roles in bearing capacity of various structures. For this reason, the regulations and design codes contain strict rules such important elements. According to studies, the failure of reinforced concrete columns and damage to columns caused by earthquake loads can be classified as follows [1]:

- Failure caused by reciprocating bending moment and low shear due to high axial pressure, which usually occurs in slender columns according to following equation:

$$\alpha = \frac{M}{Vh} = \frac{L}{2h} > 3.5 \tag{1}$$

 IntechOpen

- Failure caused by reciprocating shear and low bending moment under high axial pressure, which usually occurs in short columns according to following equation:

$$\alpha = \frac{M}{Vh} = \frac{L}{2h} < 3.5 \tag{2}$$

In Eqs. (1) and (2), M, V, h, and L refer to the bending moment, shear load, depth, and height of the column, respectively. Accordingly, the axial force is an important parameter in relation to the failure of concrete columns. One of the effective approaches to improving the axial strength as well as the ductility of reinforced concrete columns is their confinement using transverse reinforcements. The confinement limits the concrete lateral expansion and crack growth. Increases in the concrete strength and strain in the paste behavior areas are two main influences of the concrete confinement. This phenomenon leads to a decrease in the slope of the second part of the concrete strain–stress curve (its descending part) and an increase in the strain of the final stage. In this way, the area under the concrete stress–strain, which is equivalent to energy absorption, increases, in which case the ductility of the section of interest increases as well. For better understanding, **Figure 1** shows the effect of confinement on the stress–strain curve.

Building codes have set requirements for satisfying the behavior of columns during an earthquake through the amount and condition of the transverse reinforcement distribution. The failure of a column specimen due to large distances of transverse reinforcements is shown in **Figure 2**. Column failure in the previous earthquakes indicates that most of the reinforced concrete buildings were severely vulnerable to earthquakes stemming from the lack of confinement, large distances of transverse reinforcements, and other inappropriate details.

To ensure adequate ductility of reinforced concrete columns that are subjected to seismic load, ACI 318–14 building code has set the following criteria as the minimum amount of transverse reinforcements [4]. On this basis, Eqs. (3) and (4) present the minimum ratios of transverse reinforcements needed for spirals and stirrups:

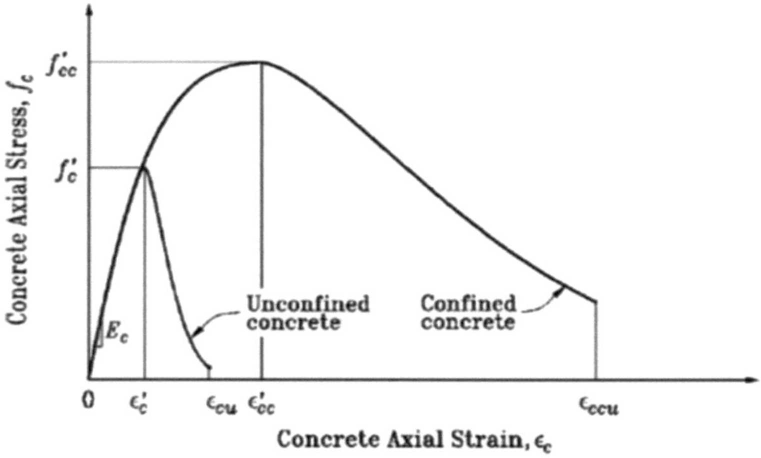

Figure 1.
The effect of confinement on the stress–strain curve [2].

Figure 2.
The severe failure in the column due to large distances of transverse reinforcements [3].

$$\rho_s \geq max \begin{cases} 0.45\left(\dfrac{A_g}{A_{Ch}} - 1\right)\dfrac{f_c}{f_{yt}} \\ 0.12\dfrac{f_c}{f_{yt}} \end{cases} \tag{3}$$

$$A_{sh}/sb_c \geq max \begin{cases} 0.3\left(\dfrac{A_g}{A_{Ch}} - 1\right)\dfrac{f_c}{f_{yt}} \\ 0.09\dfrac{f_c}{f_{yt}} \end{cases} \tag{4}$$

where A_g, A_{ch}, and A_{sh} are the cross sections of the concrete specimen, confined concrete core, and the transverse reinforcements, respectively. Moreover, f_c and f_{yt} represent the compressive strength of concrete and the yield stress of the transverse reinforcement, respectively. Finally, s and b_c denote the distance of the transverse reinforcements and the width of the confined concrete core, respectively. It is worth remaking that the fundamental principles of Eqs. (3) and (4) are based on the fact that the concrete column can preserve its strength after separating the concrete cover.

Corrosion of the transverse reinforcements is one of the important factors that not only reduces the bearing capacity but also may change the state of failure owing to the reduction of ductility. Therefore, the behavior of reinforced concrete columns affected by corrosion may be different from the original design. In this case, a full investigation of the effects of the corrosion of transverse reinforcements on the bearing capacity of reinforced concrete columns is of paramount importance. Despite some rules in building codes for decreasing the impact of corrosion on structures including reinforced concrete columns, there is not any formulation to estimate the axial strength of the corroded elements. This formulation allows civil engineers to evaluate the performance of structures and determine the axial capacity of columns under any condition and any degree of corrosion.

On the other hand, a better understanding of the effect of corrosion on the column performance helps owners and operators of reinforced concrete structures to reinforce damaged columns with a more open vision and careful strategic planning. In this chapter, one attempts to investigate the amount of reduction in bearing capacity of the axial structural elements due to the corrosion of transverse reinforcements.

2. Related works

A review of experimental studies on the reinforced concrete elements indicates that the corrosion of transverse reinforcements affects the behavior of those elements [5–9]. Shayanfar et al. [10] investigated the reduction of the compressive strength of reinforced concrete specimens under different degrees of corrosion. They subjected cubic specimens with different ratios of water-to-cement (w/c) under accelerated corrosion and then measured their compressive strength. Moreover, they proposed a formulation for the reduced compressive strength caused by corrosion. Ghanoni Bagha et al. [11] examined the reduction of compressive strength due to corrosion in their self-compacting concrete specimens using mineral admixtures. They concluded that the compressive strength of the concrete decreases by about 20% when the crack width increases by about 1 mm representing 7–12% corrosion in the reinforcements. Ahmadi et al. [12] evaluated the influence of the corrosion of transverse reinforcements on reducing the compressive strength of concrete confined by spirals. They corroded standard cylindrical specimens reinforced by spirals under the accelerated corrosion technique. The test variables included the corrosion percentage, spiral diameter, spiral pitch, and confined core diameter (concrete cover). Their experimental results demonstrated that the corrosion up to 2.5% affects reducing the confined strength. Moreover, it was observed that small spirals are less sensitive to correction. Goharrokhi et al. [13] investigated the reduction of the concrete compressive strength confined by corroded stirrups. In their experimental study, cubic reinforced concrete specimens with different diameters and spacing distances were subjected to various corrosion degrees. They showed that the reduction of the compressive strength depends on both the corrosion percentage and the stirrup diameter, for which they derived a formulation.

3. Effect of the transverse reinforcement corrosion on the axial strength of reinforced concrete elements

Ahmadi et al. [12] and Goharrokhi et al. [13] proposed Eq. (5) to calculate the compressive strength of the reinforced concrete elements using corroded transverse reinforcements:

$$f_{cc-corr} = (1 - \lambda) f_{cc} \tag{5}$$

where f_{cc} is the compressive strength of the unreinforced reinforced concrete element and $f_{cc\text{-}corr}$ denotes the compressive strength of the reinforced concrete element with corroded transverse reinforcements. Moreover, λ is the parameter for reducing the compressive strength due to corrosion of transverse reinforcements. **Figure 3** shows the fracture mechanism of the reinforced specimens with corroded spirals.

Ahmadi et al. [12] constructed cylindrical reinforced concrete specimens using spirals of the size (diameter) of 4 and 6 mm to investigate the changes in the compressive strength against different degrees of corrosion as can be seen in **Figure 4**. They expressed logarithmical formulations between the reduced compressive strength and corrosion percentage of transverse reinforcements (C_w) as follows:

$$\lambda = \frac{1}{3.3} \ln(1 + C_w) \tag{6}$$

$$\lambda = \frac{1}{1.5} \ln(1 + 0.15C_w) \tag{7}$$

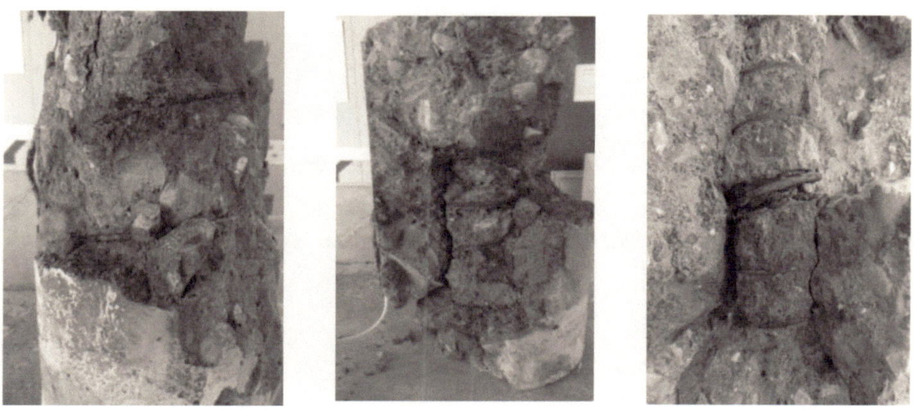

Figure 3.
The fracture mechanism of the corroded reinforced cylindrical specimens [12].

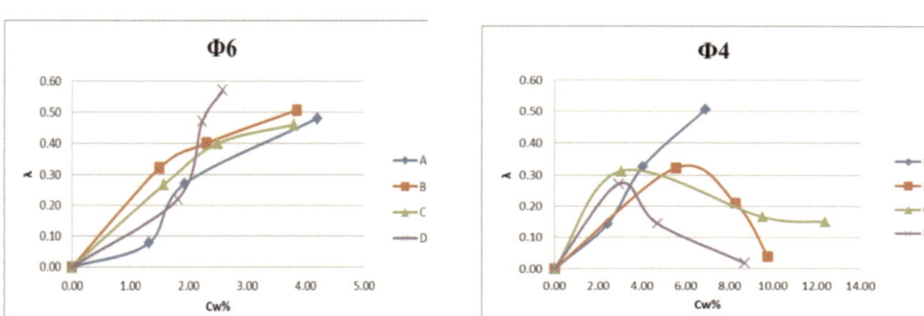

Figure 4.
Changes in the compressive strength against the corrosion degrees of under different spiral diameters [12].

Figure 5.
The fracture mechanism of the corroded reinforced rectangular specimens [13].

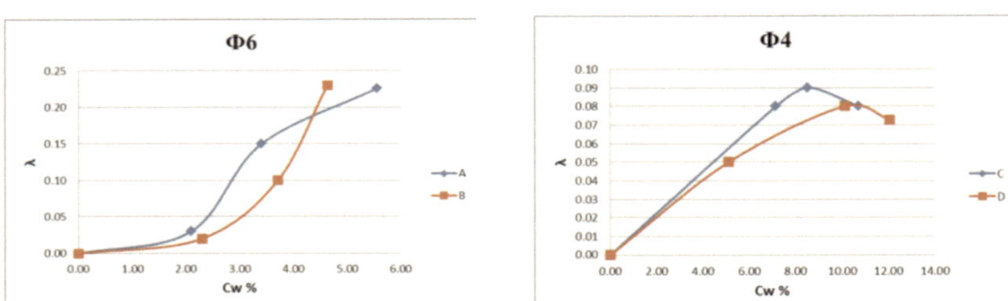

Figure 6.
Changes in the compressive strength against various corrosion degrees for different stirrup diameters [13].

Additionally, Goharrokhi et al. [13] evaluated the changes in the concrete compressive strength against different corrosion degrees for rectangular cube specimens with 4 and 6 mm diameters as can be observed in **Figures 5** and **6**. They proposed the following equations for the specimens of the diameters of 4 and 6 mm, respectively:

$$\lambda = 0.05C_w - 0.04 \tag{8}$$

$$\lambda = 0.0085C_w + 0.0066 \tag{9}$$

4. Investigation of the concrete columns with corroded transverse reinforcement

In this section, reinforced concrete columns are investigated to evaluate the reduction of the axial strength caused by the corrosion of the transverse reinforcements. For this aim, columns with circular and rectangular cross sections were selected and the problem of interest was evaluated by using the proposed formulations by Ahmadi et al. [12] and Goharrokhi et al. [13]. The circular columns, which were reinforced and evaluated by fiber reinforced polymer (FRP), include the height of 1.5 m and the diameter of 300 mm [14, 15]. In this regard, the spirals with the diameters of 4 and 6 mm placed at the intervals of 50 and 25 mm with the concrete cover were used to reinforce these columns as shown in **Figure 7(a)**. The rectangular columns evaluated by Regine and Sebastian [16] consist of the cross section of 140×140 mm and the height of 2 m. In these elements, four longitudinal reinforcements of the diameter of 8 mm were connected by the stirrups of the diameters of 4 and 6 mm with the intervals of 50 and 25 mm as can be observed in **Figure 7(b)**. The columns are named C-a-b and S-a-b, where *C* and *S* represent the circular and rectangular columns, while *a* and *b* refer to the diameter and distance of the transverse reinforcements, respectively. The characteristics of the columns as well as their strength values in the non-reinforced states are given in **Table 1**, in which f_{c0} and f_c

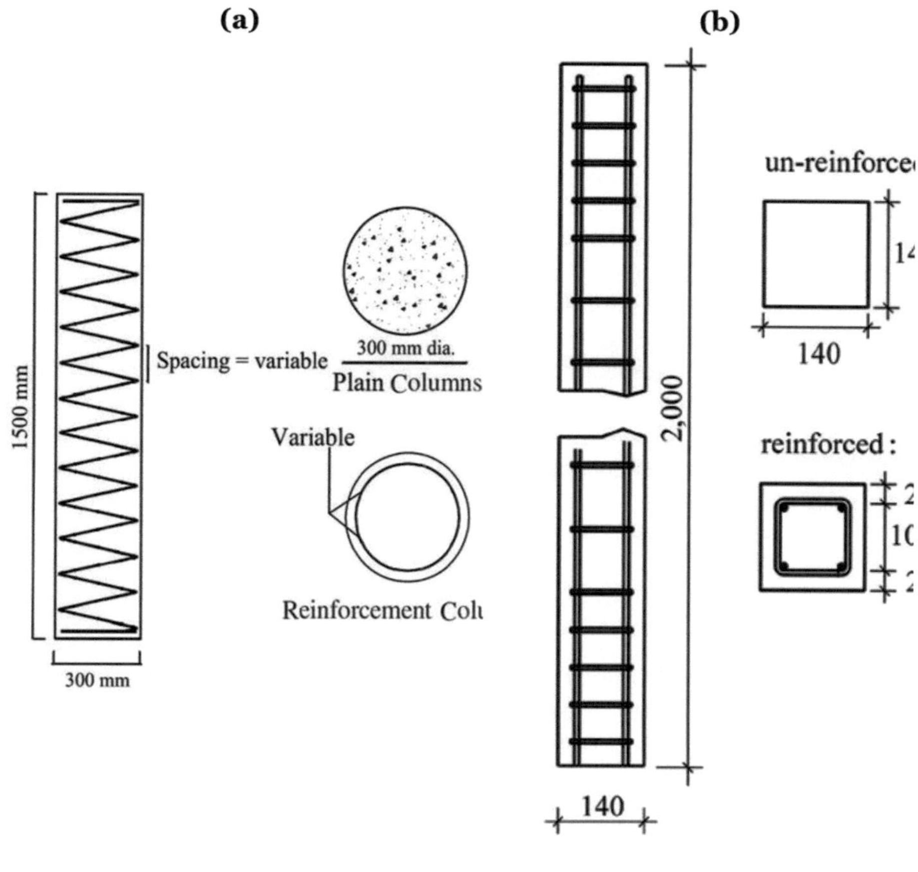

Figure 7.
The details of the axial elements: (a) the circular columns [14, 15] and (b) the rectangular columns [16].

Axial member	Dimensions (mm)	f_{c0} (MPa)	f_c (MPa)
Circular column	300x1500	42.9	34.9
Rectangular column	140x140x2000	32.4	27.2

Table 1.
Characteristics and strength of non-reinforced axial elements.

Axial member	Name	fcc/fc	$fcc_{(Cw)}/fc$				
			Cw = 2%	Cw = 4%	Cw = 6%	Cw = 8%	Cw = 10%
Circular column	C-4-50	1.06	0.87	0.73	0.60	0.50	0.41
	C-4-25	1.11	0.92	0.76	0.64	0.53	0.43
	C-6-50	1.13	0.75	0.58	0.46	0.38	0.31
	C-6-25	1.26	0.84	0.64	0.52	0.42	0.34
Rectangular column	S-4-50	1.2	1.14	1.12	1.11	1.09	1.08
	S-4-25	1.47	1.39	1.38	1.36	1.34	1.32
	S-6-50	1.47	1.35	1.2	1.06	0.92	0.77
	S-6-25	1.92	1.76	1.57	1.39	1.2	1.01

Table 2.
Strength of the columns in the corroded and conditions.

denote the strength variables of the cylindrical specimen and the non-reinforced column, respectively.

The results of the strength of the columns in the corroded and non-corroded conditions are listed in **Table 2**. On the other hand, the formulations proposed by Richart et al. [17] and Palter and Legeron [2] are applied to compute the increase in the strength obtained from the confinement of the circular sections with spirals and the rectangular sections, respectively. The main reason for utilizing their formulations originates from their approval by the ACI code. Accordingly, **Figures 8** and **9** illustrate the changes in the compressive strength caused by the corrosion of the transverse reinforcements regarding the circular and rectangular columns, respectively.

5. Conclusions

Based on the results obtained from the circular columns with spirals, the reduction of the compressive strength attributable by corrosion is initially large but it then decreases. In general, it can be concluded that the strength of spirals decreases because of corrosion, but this reduction is less at higher corrosion degrees. The results of the rectangular columns constructed by the stirrups demonstrated that the compressive strength initially decreases in the stirrups with a smaller diameter. However, the specimen with larger diameter becomes more sensitive to corrosion when the degree of corrosion increases. Therefore, it is recommended to use stirrups with small

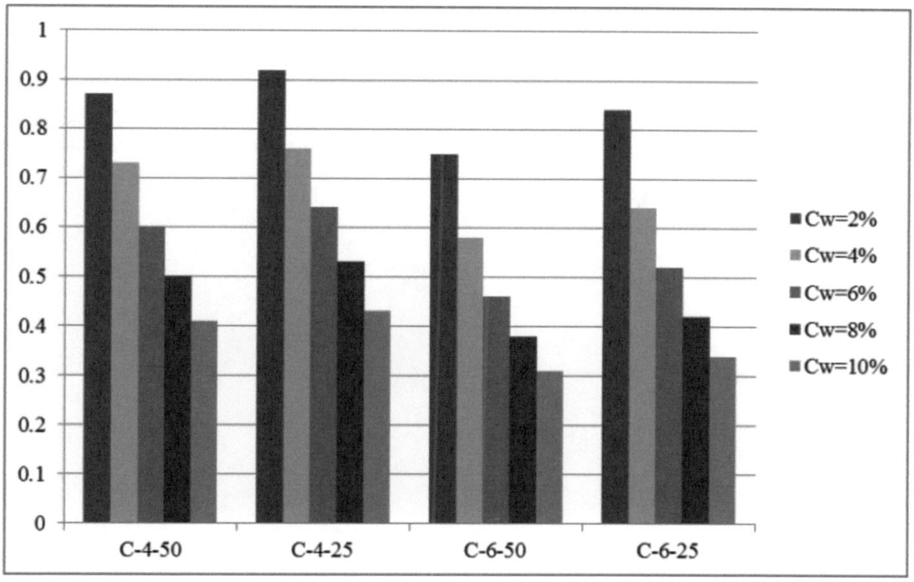

Figure 8.
Reduction in the compressive strength of the circular columns caused by the corrosion of the transverse reinforcements.

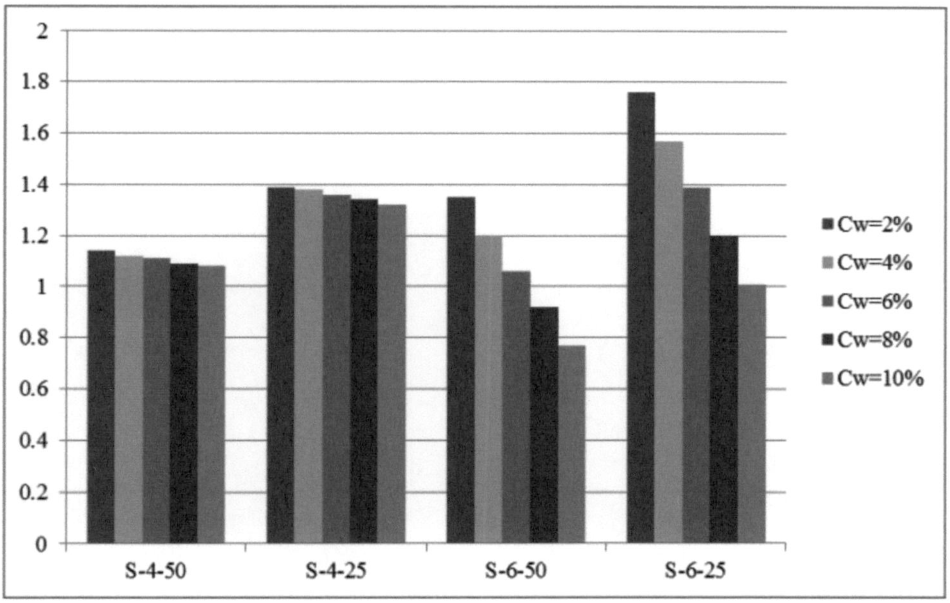

Figure 9.
Reduction in the compressive strength of the rectangular columns caused by the corrosion of the transverse reinforcements.

diameters in environments exposed to severe weather conditions. This is because such stirrups are less sensitive to high degrees of corrosion.

Generally, the corrosion of the transverse reinforcements in the circular columns reinforced by the spirals leads to a greater effect on the reduction of the strength.

Furthermore, one can conclude that the reduction in the strength of the circular columns reinforced by the spirals is larger than the corresponding rate in the rectangular columns confined by the stirrups. These cases indicate that the spirals are more sensitive to corrosion than the stirrups. Thus, it is recommended that the regulations assign more requirements and restrictions to consider the effect of the corrosion on the spirals.

Author details

Ali Goharrokhi
Civil Engineering Department, Applied Science Center of Khorasan Razavi Municipalities, Mashhad, Iran

*Address all correspondence to: goharrokhi.ali@gmail.com

References

[1] Penelis GG, Kappos AJ. Earthquake-Resistant Concrete Structures. SPON E& FN (Chapman & Hall); 1997

[2] Paultre P, Légeron F. Confinement reinforcement design for reinforced concrete columns. Journal of Structural Engineering, ASCE. 2008;**134**(5): 738-749

[3] Yoshimura K, Croston T, Kagami H, Ishiyama Y. Damage to building structures caused by the 1999 Quindío Earthquake in Colombia. 1999

[4] ACI 318 Committee. Building Code Requirements for Structural Concrete (ACI 318–14) and Commentary. American Concrete Institute. 2014

[5] Castel A, François R, Arliguie G. Mechanical behaviour of corroded reinforced concrete beams—Part 1: Experimental study of corroded beams. Materials and Structures. 2000;**33**(9): 539-544

[6] Zhang X, Li B. Seismic performance of exterior reinforced concrete beam-column joint with corroded reinforcement. Engineering Structures. 2021;**228**:111556

[7] Ou Y-C, Chen H-H. Cyclic behavior of reinforced concrete beams with corroded transverse steel reinforcement. Journal of Structural Engineering. 2014; **140**:04014050

[8] Wang XH, Liang FY. Performance of RC columns with partial length corrosion. Nuclear Engineering and Design. 2008;**238**(12):3194-3202

[9] Xia J, Jin Wl, Li Ly. Shear performance of reinforced concrete beams with corroded stirrups in chloride environment. Corrosion Science. 2011;**53**(5):1794-1805

[10] Shayanfar MA, Barkhordari MA, Ghanooni-Bagha M. Effect of longitudinal rebar corrosion on the compressive strength reduction of concrete in reinforced concrete structure. Advances in Structural Engineering. 2016

[11] Ghanooni-Bagha M, Shayanfar MA, Shirzadi-Javid AA, Ziaadiny H. Corrosion-induced reduction in compressive strength of self-compacting concretes containing mineral admixtures. Construction and Building Materials. 2016;**113**(15):221-228

[12] Ahmadi J, Shayanfar MA, Ghanooni-Bagha M, et al. An experimental investigation into the effect of transverse reinforcement corrosion on compressive strength reduction in spirally confined concrete. Iranian Journal of Science and Technology - Transactions of Civil Engineering. 2020;**44**:265-275

[13] Goharrokhi A, Ahmadi J, Shayanfar MA, et al. Effect of transverse reinforcement corrosion on compressive strength reduction of stirrup-confined concrete: An experimental study. Sādhanā. 2020;**45**:49

[14] Afifi MZ, Mohamed HM, Benmokrane B. Strength and axial behavior of circular concrete columns reinforced with CFRP bars and spirals. Journal of Composites for Construction. 2014;**18**(2):04013035

[15] Afifi MZ, Mohamed HM, Chaallal O, Benmokrane B. Confinement model for concrete columns internally confined with carbon FRP spirals and hoops. Journal of Structural Engineering. 2015; **141**(9):04014219

[16] Regine O, Sebastian O. Textile reinforced concrete for strengthening of

RC columns: A contribution to resource conservation through the preservation of structures. Construction and Building Materials. 2017;**132**:150-160

[17] Richart FE, Brandtzaeg A, Brown RL. The Failure of Plain and Spirally Reinforced Concrete in Compression. Urbana: Engineering Experiment Station, University of Illinois; 1929 (190)